BRITISH RAILW

DMUS
RAIL SYSTEMS

ELEVENTH EDITION
1998

The Complete Guide to all Diesel Multiple Units
which run on Britain's Mainline Railways
together with the Rolling Stock of
British Light Rail & Metro Systems

Peter Fox

ISBN 1 872524 98 2

© 1998. Platform 5 Publishing Ltd., 3 Wyvern House, Sark Road, Sheffield,
S2 4HG, England.

CONTENTS

Acquisition of Information 2
Organisation of Britain's Railway System 3
Introduction 6
1. Heritage DMUs 10
2. Second Generation DMUs 14
3. Diesel Electric Multiple Units 47
4. Service DMUs 50
5. DMUs Awaiting Disposal 61
6. Light Rail & Metro Systems 62
  6.1. Blackpool & Fleetwood Tramway 62
  6.2. Docklands Light Railway 65
  6.3. Greater Manchester Metrolink 67
  6.4. Midland Metro 69
  6.5. South Yorkshire Supertram 70
  6.6. Strathclyde PTE Underground 71
  6.7. Tyne & Wear Metro 72
Livery Codes 74
Owner and Operating Codes 75
Depot Codes 76

ACQUISITION OF INFORMATION

This book has been published with great difficulty. Privatisation of the railways and the splitting up of BR into different companies has been used as an excuse to deny the railway press access to official rolling stock library information, breaking a tradition of freely-supplied information which has existed for around half a century. We hope that readers will find the information accurate, but cannot be responsible for any inaccuracies.

We would like to thank the companies and individuals which have been co-operative in supplying information and would ask other companies which find this book useful to help us in future to make the book as accurate as possible.

This book is updated to 1st January 1998.

ORGANISATION OF BRITAIN'S RAILWAY SYSTEM

INFRASTRUCTURE

Britains national railway infrastructure, i.e. the track, signalling, stations and overhead line equipment is now owned by a private company called 'Railtrack'. Many stations and maintenance depots are leased to train operating companies. The exception is the infrastructure on the Isle of Wight, which is leased from Railtrack to Island Line.

DOMESTIC PASSENGER TRAIN OPERATIONS

Passenger trains are operated by train operating companies (TOCs). These TOCs operate on fixed term franchises. A list of these is appended below:

TOC	Operator	New Name
Anglia Railways	GB Trains	
Inter City East Coast	Sea Containers Ltd.	Great North Eastern Railway
Inter City West Coast	Virgin Group	Virgin Trains
Cross Country Trains	Virgin Group	Virgin Trains
Great Western Trains	Great Western Holdings	
North West Regional Railways	Great Western Holdings	North Western Trains
Midland Main Line	National Express	
Gatwick Express	National Express	
North London Railways	National Express	Silverlink
Central Trains	National Express	
Scotrail	National Express	
Merseyrail Electrics	MTL Holdings	
Regional Railways North East	MTL Holdings	
LTS Rail	Prism Rail	
South Wales & West Railway	Prism Rail	Wales & West Passenger Trains
Cardiff Railway Co.	Prism Rail	
West Anglia Great Northern	Prism Rail	
South West Trains	Stagecoach	
Island Line	Stagecoach	
Network South Central	Connex	Connex South Central
South East Trains	Connex	Connex South Eastern
Great Eastern	FirstBus	
Thameslink	GOVIA	
Chiltern Railways	M40 Trains	
Thames Trains	Victory Rail	

NOTES ON TRAIN OPERATING COMPANY OWNERS

Connex

This is a French company owned by Société Générale des Entreprises Auto-mobiles, a subsidiary of Compagnie Générale des Eaux.

FirstBus

This is a large bus company which was originally formed by the amalgamation of Badgerline and GRT bus group.

GB Trains

This is a company set up for rail privatisation.

GOVIA

A joint venture between the Go-Ahead bus company and VIA, a French public transport operating company.

Great Western Holdings

This is a jointly owned by the former Great Western Trains management. 3i plc and FirstBus.

National Express

This is an operator which runs express coach services, mainly by sub-contracting them to various bus companies. It also owns East Midlands Airport.

M40 Trains

This is owned by the former management of Chiltern Railways.

MTL Holdings

This is the former municipal bus operator Merseyside PTE which operates buses in Merseyside and London.

Prism

This is a company whose shares are owned by individuals and financial institutions. Its chairman is joint managing director of EYMS, a bus company.

Sea Containers

This is a Bermuda-based shipping company which also owns the Venice-Simplon-Orient Express.

Stagecoach

Tha largest private bus operator in the UK.

Victory Railway Holdings

This is a joint venture between the Go Ahead group and Thames Trains managament.

Virgin Group

This is the well-known company headed by Richard Branson which has interests in travel, leisure and retailing.

CHANNEL TUNNEL PASSENGER TRAIN OPERATIONS

Eurostar trains are operated by Eurostar (UK) Ltd. jointly with French Railways (SNCF) and Belgian Railways (NMBS/SNCB). Eurostar (UK) will also operate the Night Service trains jointly with SNCF, Netherlands Railways (NS) and German Railways (DB). Eurostar (UK) is now owned by a private company, London & Continental.

OWNERSHIP OF LOCOMOTIVES AND ROLLING STOCK

The locomotives of EWS and those of Eurostar are owned by those companies. Most locomotives, hauled coaching stock and multiple unit vehicles used by the passenger train operating companies are owned by three leasing companies which were originally set up by British Railways as subsidiaries and then privatised. These are:

* Forward Trust Rail (formerly Eversholt Leasing) owned by Hong Kong Shanghai Bank.
* Angel Train Contracts (owned by the Royal Bank of Scotland).
* Porterbrook Leasing Company Ltd (owned by Stagecoach Holdings).

Other vehicles are owned or operated on behalf of private owners by various private companies such as Fragonset Railways, West Coast Railway CompanyLtd., Riviera Trains Ltd. and the Venice-Simplon-Orient Express Ltd.

Further details of these companies will be found in the section on abbreviations and codes. Thus for each vehicle it is generally necessary to specify both the owner and the company which currently operates the vehicle.

A number of 'service' type vehicles are owned by Railtrack (e.g. Sandite vehicles) and others are owned by former BR Headquarters organisations which have now been privatised e.g. Serco Railtest or by railway vehicle manufacturing and repair companies. Royal Train vehicles are owned by Railtrack.

INTRODUCTION

Diesel Multiple Unit operation on Britain's main line railways has increased enormously since the end of the steam era but there have been many changes in recent years. Electrification has meant the replacement of DMUs with EMUs on many routes, whilst on other services, DMUs have replaced loco-hauled trains. Very few first generation DMUs remain and most DMU services are now operated by "Pacer", "Sprinter" or other modern air-braked Express units. A few DEMUs will be found operating on the former Southern Region, but some of these now have ex-EMU centre cars. One vehicle (71634) started life as loco-hauled coach No. 4059, was converted to an EMU trailer and is now part of DEMU 205 205!

NUMBERING

Diesel mechanical and diesel hydraulic multiple unit vehicles are numbered in the series 51000-59999. All vehicles numbered in the series 53000-53999 series were originally numbered in the series 50000-50999, and were renumbered by having 3000 added to their original numbers. All vehicles in the series 54000-54504 were originally numbered in the series 56000-56504, and were renumbered by having 2000 subtracted from their original numbers.

Diesel electric multiple unit vehicles are numbered in the series 60000-60918. A number of vehicles which were numbered in the series 60001-60100 were renumbered in 1989 to avoid conflicting with Class 60 locomotives.

DESIGN CONSIDERATIONS

Unless stated otherwise, all diesel multiple unit vehicles are of BR design, or designed by contractors for BR and have buckeye couplings and tread brakes. Seating is 3+2 or 2+2 in standard class open vehicles, 2+2 or 2+1 in first class open vehicles, 8 to a corridor standard class compartment and 6 to a corridor first class compartment.

VEHICLE CODES

The codes used by the BR Operating Department to describe the various different types of DMU vehicles and quoted in the class headings are as follows:

Diesel Mechanical & Diesel Hydraulic Units.

DMBS	Driving Motor Brake Standard
DMC	Driving Motor Composite
DMS	Driving Motor Standard
DMPMV	Driving tMotor Parcels & Miscellaneous Van
DTPMV	Driving Trailer Parcels & Miscellaneous Van
DTS	Driving Trailer Standard
MS	Motor Standard
TS	Trailer Standard

It should be noted that as all vehicles are of an open configuration the letter 'O' is omitted for all vehicles. An 'L' suffix denotes that the vehicle is fitted with a lavatory compartment. The letters (A) and (B) may be added to the above codes to differentiate between two cars of the same operating type which have differences between them. Note that a consistent system is used here, rather than the official operator codes which are sometimes inconsistent.

A composite is a vehicle containing both first and standard class accommodation, and vehicles are described as such even though most first class accommodation has now been declassified on most vehicles. This is done so as to differentiate between the different styles of seat provided in standard and erstwhile first class areas of a vehicle. At the time of writing no heritage units retained first class accommodation in use as such.

A brake vehicle is a vehicle containing seperate specific accommodation for the guard (as opposed to the use of spare driving cabs on second generation units).

Diesel Electric Units.

DMBSO Driving Motor Brake Standard (Open).
DTCsoL Driving Trailer Composite with Lavatory (Semi-Open).
DTSOL Driving Trailer Standard with Lavatory (Open).
DTSO   Driving Trailer Standard (Open).
TSO   Trailer Standard (Open).
TCsoL Trailer Composite with Lavatory (Semi-open).
TSOL   Trailer Standard with Lavatory (Open).

The notes as above apply regarding composite and brake vehicles. A semi-open composite vehicle has first class accommodation in compartments with a side corridor and standard class accommodation provided in an open saloon.

WEIGHTS & DIMENSIONS

Approximate weights in working order are given in tonnes for all vehicle types in the class headings and sub headings as appropriate.

The dimensions of each type of vehicle are given in metric units, with length followed by width. All lengths quoted are over buffers (1st generation vehicles) or couplings (2nd generation vehicles). All widths quoted are maxima.

DIAGRAMS AND DESIGN CODES

For each type of vehicle, the official design code consists of a seven character code of two letters, four numbers and another letter, e.g. DP2010A. The first five characters of this are the diagram code and are given in the class heading or sub heading. These are explained as follows:

1st Letter

This is always 'D' for a diesel multiple unit vehicle, although certain DEMUs contain EMU trailers which have an 'E' as the first letter.

2nd Letter

as follows for various vehicle types (DMMU or DHMU unless otherwise stated):

B	Driving motor passenger vehicles with a brake compartment (DEMU).
E	Driving trailer passenger vehicles (DEMU).
H	Trailer passenger vehicles without a brake compartment (DEMU).
P	Driving motor passenger vehicles without a brake compartment.
Q	Driving motor passenger vehicles with a brake compartment.
R	Non-Driving motor passenger vehicles.
S	Driving trailer passenger vehicles.
T	Trailer passenger vehicles without a brake compartment.
X	Parcels and Mails vehicles and single unit railcars.

1st Figure

This denotes the class of accommodation as follows:

2	Standard class accommodation (incl. declassified seats).
3	Composite accommodation.
5	No passenger accommodation.

2nd & 3rd Figures

These distinguish between the different designs of vehicle, each different design being allocated a unique two digit number.

Special Note

Where vehicles have been declassified the correct design code for a declassified vehicle is given, even though this may be at variance with official records which do not show the reality of the current position. A declassified composite is still referred to as a composite if it still retains the first class style seats in the erstwhile first class section of the vehicle. Its declassification is denoted by the fact that the first figure of the design code is a '2'.

ACCOMMODATION

This information is given in class headings and sub headings in the form F/S nT, where F & S denote the number of first class nd standard class seats followed by n which denotes the number of toilets. (e.g. 12/54 1T denotes 12 first class seats, 54 standard class seats and one toilet). In declassified vehicles, the capacity is still shown in terms of first and standard class seats to differentiate between the two physically different seat types available, although all seats are officially standard class in such instances.

BUILD DETAILS

LOT NUMBERS

Each batch of vehicles is allocated a Lot (or batch) number when ordered and these are quoted in class headings and sub headings.

BUILDERS

These are shown in class headings where the following abbreviations are used:

Alexander	Walter Alexander Ltd., Falkirk.
Barclay	Andrew Barclay Ltd., Kilmarnock. (now Hunslet-Barclay.)
BRCW	Birmingham Railway Carriage & Wagon Company Ltd.
Derby	BR Derby Carriage Works or British Rail Engineering Limited, Derby Carriage Works. (Became BREL Ltd. then ABB Derby and is now ADtranz Derby.)
Leyland Bus	Leyland Bus Ltd., Workington.
Metro-Cammell	Metro-Cammell Ltd., Birmingham. (now part of GEC Alsthom Ltd.)
Pressed Steel	Pressed Steel Ltd., Swindon.
York	British Rail Engineering Ltd., York. (Became BREL Ltd. and then ABB York. Now closed.)

Where a dual BR works builder is shown (e.g. Ashford/Eastleigh) the first named built the underframe and the last named built the body and assembled the vehicle. For second generation vehicles, the first name is that of the main contractor with the second name being the underframe and final assembly sub-contractor.

ABBREVIATIONS USED

The following are used throughout this section:

BSI	Bergische Stahl Industrie
DEMU	diesel electric multiple unit.
DHMU	diesel hydraulic multiple unit.
DMMU	diesel mechanical multiple unit.
DMU	diesel multiple unit (general term).
GWR	Great Western Railway
PA	Fitted with public address system.
h.p.	horsepower.
kW	Kilowatts.
T	Toilets.
TD	Toilets (suitable for disabled passengers).
m	metres.
mph	miles per hour.
r	Fitted with radio electronic token block apparatus.
t	tons.

LAYOUT

The layout in this section is as follows:

(1) Unit number.	(5) Operator code.
(2) Notes (if any).	(6) Depot code.
(3) Livery code.	(7) Individual car numbers.
(4) Owner code.	(8) Name (if any).

Thus an example of the layout is as follows:

No.	Notes	Livery	Own.	Oper.	Depot	Car 1	Car 2	Name
150 257		**RR**	P	*AR*	NC	52257	57257	Queen Boadicea

For off-loan or stored vehicles, the last storage location is given where known.

1. 'HERITAGE' DIESEL MULTIPLE UNITS

Very few first generation diesel multiple units remain. These are now referred to as 'heritage' units. Standard features are as follows:

Brakes:

All units are vacuum braked.

Lighting:

All cars are now fitted with fluorescent lighting.

Couplings:

Screw couplings are used on all vehicles. All remaining first generation vehicles may be coupled together to work in multiple up to a maximum of 6 motor cars or 12 cars in total in a formation. First generation vehicles may not be coupled in multiple with second generation vehicles.

CLASS 101   METRO-CAMMELL

Engines: Two Leyland of 112 kW (150 h.p.) per power car.
Transmission: Mechanical. Cardan shaft and freewheel to a four-speed epicyclic gearbox with a further cardan shaft to the final drive, each engine driving the inner axle of one bogie.
Gangways: Midland scissors type. Within unit only.
Doors: Slam.
Bogies: DD15 (motor) and DT11 (trailer).
Dimensions: 18.49 x 2.82 m.
Seats: 3+2 facing (2+2 in first class).

51175–51253. DMBS. Dia. DQ202. Lot No. 30467 1958–59. –/52. 32.5 t.
51426–51463. DMBS. Dia. DQ202. Lot No. 30500 1959. –/52. (–/49 with additional luggage rack for Gatwick sets – Dia. DQ232) 32.5 t.
53164. DMBS. Dia. DQ202. Lot No. 30254 1956. –/52. 32.5 t.
53198–53204. DMBS. Dia. DQ202. Lot No. 30259 1957. –/52. 32.5 t.
53211–53228. DMBS. Dia. DQ202. Lot No. 30261 1957. –/52. 32.5 t.
53253–53256. DMBS. Dia. DQ202. Lot No. 30266 1957. –/52. 32.5 t.
53311–53314. DMBS. Dia. DQ202. Lot No. 30275 1958. –/52. (–/49 with additional luggage rack for Gatwick sets – Dia. DQ232) 32.5 t.
51496–51533. DMCL or DMSL. Dia. DP317 or DP210. Lot No. 30501 1959. 12/46 1T with additional luggage racks. 32.5t.
51800. DMBS. Dia. DQ202. Lot No. 30587 1956. –/52. 32.5 t.
51803. DMSL. Dia. DP210. Lot No. 30588 1959. –/72 1T. 32.5 t.
53160–53163. DMSL. Dia. DP214. Lot No. 30253 1956. –/72 1T. 32.5 t.
53170–53171. DMSL. Dia. DP214. Lot No. 30255 1957. –/72 1T. 32.5 t.
53177. DMSL. Dia. DP214. Lot No. 30256 1957. –/72 1T. 32.5 t.
53266–53269. DMSL. Dia. DP210. Lot No. 30267 1957. –/72 1T. 32.5 t.
53322–53327. DMCL. Dia. DP317. Lot No. 30276 1958. 12/46 1T with additional luggage racks. 32.5 t.

53746. DMSL. Dia. DP210. Lot No. 30271 1957. –/72 1T. 32.5 t.
54055–54061. DTSL. DS206. Lot No. 30260 1957. –/72 1T. 25.5 t.
54062–54091. DTSL. Dia. DS206. Lot No. 30262 1957. –/72 1T. 25.5 t.
54343–54408. DTSL. Dia. DS206. Lot No. 30468 1958. –/72 1T. 25.5 t.
59303. TSL. Dia. DT202. Lot No. 30273 1957. –/71 1T. 25.5 t.
59539. TSL. Dia. DT228. Lot No. 30502 1957. –/72 1T. 25.5 t.

Refurbished 2-car Sets. DMBS–DTSL.

101 651	**RR**	A	*NW*	LO (U)	53201	54379
101 652	**RR**	A	*NW*	LO (U)	53198	54346
101 653	**RR**	A	*NW*	LO	51426	54358
101 654	**RR**	A	*NW*	LO	51800	54408
101 655	**RR**	A	*NW*	LO	51428	54062
101 656	**RR**	A	*NW*	LO	51230	54056
101 657	**RR**	A	*NW*	LO	53211	54085
101 658	**RR**	A	*NW*	LO	51175	54091
101 659	**RR**	A	*NW*	LO	51213	54352
101 660	**RR**	A	*NW*	LO	51189	54343
101 661	**RR**	A	*NW*	LO	51463	54365
101 662	**RR**	A	*NW*	LO	53228	54055
101 663	**RR**	A	*NW*	LO	51201	54347
101 664	**RR**	A	*NW*	LO	51442	54061
101 665	**RR**	A	*NW*	LO	51429	54393

Refurbished Twin Power Car and 3-Car Sets. DMBS–DMSL or DMBS–TSL–DMSL.

Non-Standard livery: Caledonian Blue.

101 676	**RR**	A	*NW*	LO	51205		51803
101 677	**RR**	A	*NW*	LO	51179		51496
101 678	**RR**	A	*NW*	LO	51210		53746
101 679	**RR**	A	*NW*	LO	51224		51533
101 680	**RR**	A	*NW*	LO	53204		53163
101 681	**RR**	A	*NW*	LO	51228		51506
101 682	**RR**	A	*NW*	LO	53256		51505
101 683	**RR**	A	*NW*	LO	51177	59303	53269
101 684	**S**	A	*SR*	CK	51187		51509
101 685	**G**	A	*NW*	LO	53164	59539	53160
101 686	**S**	A	*SR*	CK	51231		51500
101 687	**S**	A	*SR*	CK	51247		51512
101 688	**S**	A	*SR*	CK	51431		51501
101 689	**S**	A	*SR*	CK	51185		51511
101 690	**S**	A	*SR*	CK	51435		53177
101 691	**S**	A	*SR*	CK	51253		53171
101 692	**0**	A	*SR*	CK	53253		53170
101 693	**S**	A	*SR*	CK	51192		53266
101 694	**S**	A	*SR*	CK	51188		53268
101 695	**S**	A	*SR*	CK	51226		51499

Note: The trailers of units 101 683 and 101 685 are normally removed for the winter period and stored at Chester CSD.

Unrefurbished Twin Power Car Sets. DMBS–DMCL. These sets have seats removed and additional luggage racks. These modifications were carried out when they were used on Reading–Gatwick Airport services.

Note: Some units show 'L' instead of the official class prefix.

101 835	**RR**	A	*NW*	LO	51432	51498
101 840	**N**	A	*NW*	LO	53311	53322
101 842	**N**	A		KI	53314	53327

CLASS 117   PRESSED STEEL SUBURBAN

DMBS–TSL–DMS (refurbished) or DMBS–DMS (facelifted).
Engines: Two Leyland 680/1 of 112 kW (150 h.p.) per power car.
Transmission: Mechanical. Cardan shaft and freewheel to a four-speed epicyclic gearbox with a further cardan shaft to the final drive, each engine driving the inner axle of one bogie.
Gangways: GWR suspension type. Within unit only.
Bogies: DD10 (motor) and DT9 (trailer).
Dimensions: 20.45 x 2.82 m.
Seats: 3+2 facing.

DMBS. Dia. DQ220. Lot No. 30546 1959–60. –/65. 36.5 t.
TSL. Dia. DT230. Lot No. 30547 1959–60. –/78 2T. 30.5 t.
DMS. Dia. DP221. Lot No. 30548 1959–60. –/89. 36.5 t.

Notes: Some units show 'L' instead of the official class prefix.

117 301	**RR**	A	*SR*	HA	51353	59505	51395	
117 306	**RR**	A	*SR*	HA	51369	59521	51411	
117 308	**RR**	A	*SR*	HA	51371	59509	51413	
117 310	**RR**	A	*SR*	HA	51373	59486	51381	
117 311	**RR**	A	*SR*	HA	51334	59500	51376	
117 313	**RR**	A	*SR*	HA	51339	59492	51382	
117 314	**RR**	A	*SR*	HA (S)	51352	59489	51394	
117 700	**N**	A	*NL*	BY	51332		51374	
117 701	**N**	A	*NL*	BY	51350		51392	Marston Vale
117 702	**N**	A	*NL*	BY	51356		51398	
117 703	**N**	A		BY	51359		51401	
117 704	**N**	A	*NL*	BY	51341		51383	
117 705	**N**	A	*NL*	BY	51358		51400	
117 706	**N**	A	*NL*	BY	51366		51408	
117 707	**N**	A	*NL*	BY	51335		51377	
117 708	**N**	A		KI	51336		51378	
117 709	**N**	A		KI	51344		51386	
117 720	**N**	A	*NL*	BY	51354		51396	
117 721	**N**	A	*NL*	BY	51363		51405	
117 724	**N**	A	*NL*	BY	51333		51375	

Names:

51332 of 117 700 is named 'Marston Vale'.
51358 of 117 705 is named 'LESLIE CRABBE'.

CLASS 121   PRESSED STEEL SUBURBAN

DMBS.
Engines: Two Leyland 1595 of 112 kW (150 h.p.) per power car.
Transmission: Mechanical. Cardan shaft and freewheel to a four-speed epicyclic gearbox with a further cardan shaft to the final drive, each engine driving the inner axle of one bogie.
Gangways: Non gangwayed single cars with cabs at each end.
Bogies: DD10 (motor) and DT9 (trailer).
Dimensions: 20.45 x 2.82 m.
Seats: 3+2 facing.

DMBS. Dia. DX201. Lot No. 30518 1960. –/65. 38.0 t.

Note: Some of the sets show 'L' instead of the official class prefix.

121 127	**N**	A	*GE*	IL	55027
121 129	**N**	A	*GE*	IL	55029
121 131	**N**	A	*GE*	IL	55031

CLASS 122   GLOUCESTER SUBURBAN

DMBS.
Engines: Two AEC 220 of 112 kW (150 h.p.).
Transmission: Mechanical. Cardan shaft and freewheel to a four-speed epicyclic gearbox with a further cardan shaft to the final drive, each engine driving the inner axle of one bogie.
Gangways: Non gangwayed single car with cabs at each end.
Bogies: DD10 (motor) and DT9 (trailer).
Dimensions: 20.45 x 2.82 m.
Seats: 3+2 facing.

DMBS. Dia. DX202. Lot No. 30419 1958. 36.5 t.

Note: This unit is used as a crew-training vehicle and does not carry its unit number.

| 122 012 | **LH** | E | *CR* | TE | 55012 |

2. SECOND GENERATION DMUS

Unit Types

There are six basic types of second generation vehicle as referred to in the class headings as follows:

Pacers (Railbuses). Folding power operated exterior doors. Bus-type 3+2 (2+2 on class 141) largely unidirectional seating. Limited luggage space. Four wheel chassis. 75 m.p.h.

Sprinter. Sliding power operated exterior double doors to large entrance vestibules. High backed 3+2 seating. Limited luggage space. 75 m.p.h.

Super·Sprinter. Sliding/sliding plug power-operated exterior doors. High backed 2+2 largely unidirectional seating with some tables. 75 m.p.h.

Express. Sliding plug power-operated exterior doors. Air conditioned. High backed 2+2 half-facing and half-unidirectional seating with some tables. 90 m.p.h.

Network Turbo. Sliding power operated exterior double doors to large entrance vestibules. 3+2 seating. Limited luggage space. 75 or 90 m.p.h.

Turbostar. Sliding power operated exterior double doors to large entrance vestibules. 2+2 seating. Limited luggage space. 100 m.p.h.

Public Address System: All vehicles are equipped with public address, with transmission equipment on driving vehicles.

Gangways: Unless stated otherwise, all vehicles have flexible diaphragm gangways.

Couplings: Unless otherwise stated all vehicles are fitted with BSI automatic couplings at their outer ends. Railbus types are fitted with bar couplings at their inner ends, but all other types have BSI couplings at their inner ends unless otherwise stated.

Brakes: All vehicles are fitted with electro-pneumatic and air brakes.

CLASS 141 — LEYLAND BUS/BREL RAILBUS

DMS–DMSL. Built from Leyland National bus parts on four-wheeled underframes.

Engines: One Leyland TL11 152 kW (205 h.p.) (* Cummins LT10-R) per car.
Transmission: Hydraulic. Voith T211r with Gmeinder final drive.
Gangways: Within unit only.
Doors: Folding.
Dimensions: 15.45 x 2.50 m.
Accommodation: 2+2 bus style.
Maximum Speed: 75 m.p.h.

DMS. Dia. DP228 Lot No. 30977 Derby 1984. Modified by Barclay 1988–89. –/50. 26.0 t.
DMSL. Dia. DP229 Lot No. 30978 Derby 1984. Modified by Barclay 1988–89. –/44 1T. 26.5 t.

141 101		**Y**	P	*NE*	NL (S)	55521 55541
141 102		**Y**	P	*NE*	HT (S)	55502 55522
141 103		**Y**	P		ZB	55503 55523
141 105		**Y**	P		ZB	55505 55525
141 106		**Y**	P		ZB	55506 55526
141 107		**Y**	P		ZB	55507 55527
141 108		**Y**	P		ZB	55508 55528
141 109		**Y**	P	*NE*	HT (S)	55509 55529
141 110		**Y**	P		ZB	55510 55530
141 111		**Y**	P	*NE*	HT (S)	55511 55531
141 112		**Y**	P		ZB	55512 55532
141 113	*	**Y**	P	*NE*	NL	55513 55533
141 114		**Y**	P	*NE*	NL (S)	55514 55534
141 115		**Y**	P	*NE*	HT (S)	55515 55535
141 116		**Y**	P		ZB	55516 55536
141 117		**Y**	P	*NE*	HT (S)	55517 55537
141 118		**Y**	P		ZB	55518 55538
141 119		**Y**	P	*NE*	NL	55519 55539
141 120		**Y**	P		ZB	55520 55540

CLASS 142 — LEYLAND BUS/BREL RAILBUS

DMS–DMSL. Development of Class 141 with wider body and improved appearance.

Engines: One Cummins LTA10-R of 170 kW (225 h.p.) per car.
Transmission: Hydraulic. Voith T211r with Gmeinder final drive.
Gangways: Within unit only.
Doors: Folding.
Dimensions: 15.55 x 2.80 m.
Accommodation: 2+3 bus style.
Maximum Speed: 75 m.p.h.

Non-Standard Livery: Chocolate & Cream.

55542–55591. DMS. Dia. DP234 Lot No. 31003 Derby 1985–6. –/62. 24.5 t.
55592–55641. DMSL. Dia. DP235 Lot No. 31004 Derby 1985–6. –/59 1T. 25.0 t.

55701–55746. DMS. Dia. DP234 Lot No. 31013 Derby 1986–7. –/62. 24.5 t.
55747–55792. DMSL. Dia. DP235 Lot No. 31014 Derby 1986–7. –/59 1T. 25.0 t.

142 001	**GM**	A	*NW*	NH	55542	55592
142 002	**GM**	A	*NW*	NH	55543	55593
142 003	**GM**	A	*NW*	NH	55544	55594
142 004	**GM**	A	*NW*	NH	55545	55595
142 005	**GM**	A	*NW*	NH	55546	55596
142 006	**GM**	A	*NW*	NH	55547	55597
142 007	**GM**	A	*NW*	NH	55548	55598
142 008	**GM**	A	*NW*	NH	55549	55599
142 009	**GM**	A	*NW*	NH	55550	55600
142 010	**GM**	A	*NW*	NH	55551	55601
142 011	**GM**	A	*NW*	NH	55552	55602
142 012	**GM**	A	*NW*	NH	55553	55603
142 013	**GM**	A	*NW*	NH	55554	55604
142 014	**GM**	A	*NW*	NH	55555	55605
142 015	**RR**	A	*NE*	HT	55556	55606
142 016	**RR**	A	*NE*	HT	55557	55607
142 017	**T**	A	*NE*	HT	55558	55608
142 018	**T**	A	*NE*	HT	55559	55609
142 019	**T**	A	*NE*	HT	55560	55610
142 020	**T**	A	*NE*	HT	55561	55611
142 021	**T**	A	*NE*	HT	55562	55612
142 022	**T**	A	*NE*	HT	55563	55613
142 023	**RR**	A	*NW*	NH	55564	55614
142 024	**RR**	A	*NE*	HT	55565	55615
142 025	**O**	A	*NE*	HT	55566	55616
142 026	**O**	A	*NE*	HT	55567	55617
142 027	**GM**	A	*NW*	NH	55568	55618
142 028	**GM**	A	*NW*	NH	55569	55619
142 029	**GM**	A	*NW*	NH	55570	55620
142 030	**GM**	A	*NW*	NH	55571	55621
142 031	**GM**	A	*NW*	NH	55572	55622
142 032	**GM**	A	*NW*	NH	55573	55623
142 033	**RR**	A	*NW*	NH	55574	55624
142 034	**GM**	A	*NW*	NH	55575	55625
142 035	**GM**	A	*NW*	NH	55576	55626
142 036	**RR**	A	*NW*	NH	55577	55627
142 037	**GM**	A	*NW*	NH	55578	55628
142 038	**GM**	A	*NW*	NH	55579	55629
142 039	**GM**	A	*NW*	NH	55580	55630
142 040	**GM**	A	*NW*	NH	55581	55631
142 041	**GM**	A	*NW*	NH	55582	55632
142 042	**GM**	A	*NW*	NH	55583	55633
142 043	**GM**	A	*NW*	NH	55584	55634
142 044	**RR**	A	*NW*	NH	55585	55635

▲ 3-car Class 101 No. 101 683 waits at Blaneau Ffestiniog on 15th September 1997 before working the 17.27 to Llandudno. Regional Railways livery is carried. **Peter Fox**

▼ Network SouthEast liveried Class 117 No. 117 706 pauses at Ridgemont, Bedfordshire on 1st October 1996 with the 12.40 Bedford–Bletchley. **Kevin Conkey**

▲ Class 121 'Bubble Cars' Nos. L127 and L129 at Colchester depot on 21st September 1997. These units are now based here for use on the Sudbury branch.
Michael J. Collins

▼ The only Class 122 still numbered in the capital stock series is No. 55012 and is in fact used as a crew-training vehicle. The unit, which carries Loadhaul livery, is pictured here with an Immingham to Landor Street run on 2nd May 1997.
Bob Sweet

▲ There are now only two Class 141 units in traffic. One of them No. 141 113 has Voith transmission and is pictured here near Doncaster with the 13.44 Scunthorpe–Doncaster on 20th January 1997. The unit is in West Yorkshire PTE livery. **Gary Pierrepont**

▼ Merseytravel liveried Class 142 No. 142 058 at Siddick on the Solway coast on 8th March 1997 with the 10.15 Whitehaven–Carlisle. **Dave McAlone**

▲ Class 143 No. 143 608 passes Burton-on-Trent running ECS to Derby on 12th June 1996.
Hugh Ballantyne

▼ Class 144 No. 144 008 pauses at Milford Sidings with a crew-training run on 10th October 1996.
Nic Joynson

Centro Trains liveried Class 150/1 No. 150 016 departs from Hatton with the 11.00 Leamington Spa–Great Malvern service on 29th February 1996.

Hugh Ballantyne

▲ Greater Manchester PTE liveried Class 150/2 No. 150 224 en-route between Manchester Oxford Road and Piccadilly on 28th May 1997.　**Hugh Ballantyne**

▼ Class 153 No. 153 324 pauses at Bamber Bridge on 17th June 1997 with the 19.09 Colne–Blackpool South service.　**Martyn Hilbert**

Class 155 No. 155 342 passes the Rochdale Canal near Castleton with the 07.40 Selby–Manchester Victoria service during May 1997.

Vincent Eastwood

▲ Class 156 No. 156 425 pauses at Lostock Hall with a Sundays only 08.39 Blackpool North–Carlisle 'Dalesrail' service. The unit is painted in unbranded North West Regional Railways livery. **Martyn Hilbert**

▼ Class 158/0 No. 158 765 passes Heeley, Sheffield on 1st August 1997 with the 16.12 Cleethorpes–Manchester Airport service. The unit carries Regional Railways Express livery. **P. Renard**

142 045	GM	A	NW	NH	55586	55636
142 046	GM	A	NW	NH	55587	55637
142 047	RR	A	NW	NH	55588	55638
142 048	RR	A	NW	NH	55589	55639
142 049	GM	A	NW	NH	55590	55640
142 050	PR	A	NE	HT	55591	55641
142 051	MT	A	NW	NH	55701	55747
142 052	MT	A	NW	NH	55702	55748
142 053	MT	A	NW	NH	55703	55749
142 054	MT	A	NW	NH	55704	55750
142 055	MT	A	NW	NH	55705	55751
142 056	MT	A	NW	NH	55706	55752
142 057	MT	A	NW	NH	55707	55753
142 058	MT	A	NW	NH	55708	55754
142 060	GM	A	NW	NH	55710	55756
142 061	GM	A	NW	NH	55711	55757
142 062	GM	A	NW	NH	55712	55758
142 063	GM	A	NW	NH	55713	55759
142 064	GM	A	NW	NH	55714	55760
142 065	PR	A	NE	HT	55715	55761
142 066	PR	A	NE	HT	55716	55762
142 067	GM	A	NW	NH	55717	55763
142 068	GM	A	NW	NH	55718	55764
142 069	GM	A	NW	NH	55719	55765
142 070	GM	A	NW	NH	55720	55766
142 071	RR	A	NE	NL	55721	55767
142 072	RR	A	NE	NL	55722	55768
142 073	RR	A	NE	NL	55723	55769
142 074	RR	A	NE	NL	55724	55770
142 075	RR	A	NE	NL	55725	55771
142 076	RR	A	NE	NL	55726	55772
142 077	RR	A	NE	NL	55727	55773
142 078	RR	A	NE	NL	55728	55774
142 079	RR	A	NE	NL	55729	55775
142 080	RR	A	NE	NL	55730	55776
142 081	RR	A	NE	NL	55731	55777
142 082	RR	A	NE	NL	55732	55778
142 083	RR	A	NE	NL	55733	55779
142 084	RR	A	NE	NL	55734	55780
142 085	RR	A	NE	NL	55735	55781
142 086	RR	A	NE	NL	55736	55782
142 087	RR	A	NE	NL	55737	55783
142 088	RR	A	NE	NL	55738	55784
142 089	RR	A	NE	NL	55739	55785
142 090	RR	A	NE	NL	55740	55786
142 091	RR	A	NE	NL	55741	55787
142 092	RR	A	NE	NL	55742	55788
142 093	RR	A	NE	NL	55743	55789
142 094	RR	A	NE	NL	55744	55790
142 095	RR	A	NE	NL	55745	55791
142 096	RR	A	NE	NL	55746	55792

CLASS 143   ALEXANDER/BARCLAY RAILBUS

DMS–DMSL. Similar design to Class 142, but bodies built by W. Alexander with Barclay underframes.

Engines: One Cummins LTA10-R of 170 kW (225 h.p.) per car.
Transmission: Hydraulic. Voith T211r with Gmeinder final drive.
Gangways: Within unit only.
Doors: Folding.
Dimensions: 15.55 x 2.70 m.
Accommodation: 2+3 bus style.
Maximum Speed: 75 m.p.h.

DMS. Dia. DP236 Lot No. 31005 Andrew Barclay 1985–6. –/62. 24.5 t.
DMSL. Dia. DP237 Lot No. 31006 Andrew Barclay 1985–6. –/60 1T. 25.0 t.

Note: 143 601/10/4 are owned by Mid-Glamorgan County Council, 143 609 is owned by South Glamorgan County Council and 143 617–9 are owned by West Glamorgan County Council although managed by Porterbrook Leasing Company.

g Fitted with global positioning system.

143 601		**RR**	P	*WW*	CF	55642	55667	
143 602	g	**RR**	P	*CA*	CF	55651	55668	
143 603	g	**RR**	P	*CA*	CF	55658	55669	
143 604	g	**RR**	P	*CA*	CF	55645	55670	
143 605	g	**RR**	P	*CA*	CF	55646	55671	
143 606	g	**RR**	P	*CA*	CF	55647	55672	
143 607	g	**RR**	P	*CA*	CF	55648	55673	
143 608	g	**RR**	P	*CA*	CF	55649	55674	
143 609	g	**RR**	P	*CA*	CF	55650	55675	
143 610		**RR**	P	*WW*	CF	55643	55676	
143 611	g	**RR**	P	*CA*	CF	55652	55677	
143 612		**RR**	P	*WW*	CF	55653	55678	
143 613	g	**RR**	P	*CA*	CF	55654	55679	
143 614		**RR**	P	*WW*	CF	55655	55680	
143 615	g	**RR**	P	*CA*	CF	55656	55681	
143 616	g	**RR**	P	*CA*	CF	55657	55682	
143 617		**RR**	P	*WW*	CF	55644	55683	Bewick's Swan
143 618		**RR**	P	*WW*	CF	55659	55684	Mute Swan
143 619		**RR**	P	*WW*	CF	55660	55685	Whooper Swan
143 620		**RR**	P	*WW*	CF	55661	55686	
143 621		**RR**	P	*WW*	CF	55662	55687	
143 622		**RR**	P	*WW*	CF	55663	55688	
143 623		**RR**	P	*WW*	CF	55664	55689	
143 624		**RR**	P	*WW*	CF	55665	55690	
143 625		**RR**	P	*WW*	CF	55666	55691	

CLASS 144   ALEXANDER/BREL RAILBUS

DMS–DMSL or DMS–MS–DMSL. Similar design to Class 143, but under-frames built by BREL as subcontractor to W. Alexander.

Engines: One Cummins LTA10-R of 170 kW (225 h.p.) per car.
Transmission: Hydraulic. Voith T211r with Gmeinder final drive.
Gangways: Within unit only.
Doors: Folding.
Dimensions: 15.25 x 2.70 m.
Accommodation: 2+3 bus style.
Maximum Speed: 75 m.p.h.

DMS. Dia. DP240 Lot No. 31015 Derby 1986–7. –/62 and wheelchair space. 24.2 t.
MS. Dia. DR205 Lot No. Derby 31037 1987. –/73. 22.6 t.
DMSL. Dia. DP241 Lot No. Derby 31016 1986–7. –/60 1T. 25.0 t.

Note: The centre cars of the three-car units are owned by West Yorkshire PTE, although managed by Porterbrook Leasing Company.

144 001	**Y**	P	*NE*	NL	55801		55824
144 002	**Y**	P	*NE*	NL	55802		55825
144 003	**Y**	P	*NE*	NL	55803		55826
144 004	**Y**	P	*NE*	NL	55804		55827
144 005	**Y**	P	*NE*	NL	55805		55828
144 006	**Y**	P	*NE*	NL	55806		55829
144 007	**Y**	P	*NE*	NL	55807		55830
144 008	**Y**	P	*NE*	NL	55808		55831
144 009	**Y**	P	*NE*	NL	55809		55832
144 010	**Y**	P	*NE*	NL	55810		55833
144 011	**RR**	P	*NE*	NL	55811		55834
144 012	**RR**	P	*NE*	NL	55812		55835
144 013	**RR**	P	*NE*	NL	55813		55836
144 014	**Y**	P	*NE*	NL	55814	55850	55837
144 015	**Y**	P	*NE*	NL	55815	55851	55838
144 016	**Y**	P	*NE*	NL	55816	55852	55839
144 017	**Y**	P	*NE*	NL	55817	55853	55840
144 018	**Y**	P	*NE*	NL	55818	55854	55841
144 019	**Y**	P	*NE*	NL	55819	55855	55842
144 020	**Y**	P	*NE*	NL	55820	55856	55843
144 021	**Y**	P	*NE*	NL	55821	55857	55844
144 022	**Y**	P	*NE*	NL	55822	55858	55845
144 023	**Y**	P	*NE*	NL	55823	55859	55846

CLASS 150/0   BREL PROTOTYPE SPRINTER

DMSL–MS–DMS. Prototype Sprinter.

Engines: One Cummins NT855R5 of 210 kW (285 h.p.) per car.
Transmission: Hydraulic. Voith T211r with Gmeinder final drive.

Bogies: One BX8P and one BX8T.
Couplings: BSI at outer end of driving vehicles, bar non-driving ends.
Gangways: Within unit only.
Doors: Sliding.
Accommodation: 2+3 (mainly unidirectional).
Dimensions: 20.06 x 2.82 m (outer cars), 20.18 x 2.82 m (inner car).
Maximum Speed: 75 m.p.h.

DMSL. Dia. DP230. Lot No. 30984 York 1984. –/72 1T. 35.8 t.
MS. Dia. DR202. Lot No. 30986 York 1984. –/92. 34.4 t.
DMS. Dia. DP231. Lot No. 30985 York 1984. –/76. 35.6 t.

Note: 150 002 was converted to 154 002 at RTC Derby in 1986, but was later converted back to a Class 150.

150 001	**CE**	A	*CT*	TS	55200	55400	55300
150 002	**CE**	A	*CT*	TS	55201	55401	55301

CLASS 150/1   BREL SPRINTER

DMSL–DMS or DMSL–DMSL–DMS or DMSL–DMS–DMS.

Engines: One Cummins NT855R5 of 210 kW (285 h.p.) per car.
Transmission: Hydraulic. Voith T211r with Gmeinder final drive.
Bogies: One BP38 and one BT38.
Gangways: Within unit only.
Doors: Sliding.
Accommodation: 2+3 facing. († Reseated with part unidirectional seating and part facing).
Dimensions: 20.06 x 2.82 m.
Maximum Speed: 75 m.p.h.

DMSL. Dia. DP238. Lot No. 31011 York 1985–6. –/68 1T (–/64 1T*, –/72 1T†). 36.5 t.
DMS. Dia. DP239. Lot No. 31012 York 1985–6. –/70 (–/76†, –/66§). 38.45 t.

Note: The centre cars of three-car units are Class 150/2 vehicles. For details see next Class.

150 010	r† **CE**	A	*CT*	TS	52110	57226	57110
150 011	r† **CE**	A	*CT*	TS	52111	57206	57111
150 012	r† **CE**	A	*CT*	TS	52112	52204	57112
150 013	r† **CE**	A	*CT*	TS	52113	52226	57113
150 014	r† **CE**	A	*CT*	TS	52114	57204	57114
150 015	r† **CE**	A	*CT*	TS	52115	52206	57115
150 016	r† **CE**	A	*CT*	TS	52116	57212	57116
150 021	r† **CE**	A	*CT*	TS	52121	57220	57121
150 101	r† **CE**	A	*CT*	TS	52101		57101
150 102	r† **CE**	A	*CT*	TS	52102		57102
150 103	r† **CE**	A	*CT*	TS	52103		57103
150 104	r† **CE**	A	*CT*	TS	52104		57104
150 105	r† **CE**	A	*CT*	TS	52105		57105
150 106	r† **CE**	A	*CT*	TS	52106		57106

150 107	r† **CE**	A	*CT*	TS	52107	57107
150 108	r† **CE**	A	*CT*	TS	52108	57108
150 109	r† **CE**	A	*CT*	TS	52109	57109
150 118	r† **CE**	A	*CT*	TS	52118	57118
150 119	r† **CE**	A	*CT*	TS	52119	57119
150 120	r† **CE**	A	*CT*	TS	52120	57120
150 121	r† **CE**	A	*CT*	TS	52121	57121
150 122	r† **CE**	A	*CT*	TS	52122	57122
150 123	r† **CE**	A	*CT*	TS	52123	57123
150 124	r† **CE**	A	*CT*	TS	52124	57124
150 125	r† **CE**	A	*CT*	TS	52125	57125
150 126	r† **CE**	A	*CT*	TS	52126	57126
150 127	r† **CE**	A	*CT*	TS	52127	57127
150 128	r† **CE**	A	*CT*	TS	52128	57128
150 129	r† **CE**	A	*CT*	TS	52129	57129
150 130	r† **CE**	A	*CT*	TS	52130	57130
150 131	r† **CE**	A	*CT*	TS	52131	57131
150 132	r† **CE**	A	*CT*	TS	52132	57132
150 133	r* **GM**	A	*NW*	NH	52133	57133
150 134	r* **GM**	A	*NW*	NH	52134	57134
150 135	r* **GM**	A	*NW*	NH	52135	57135
150 136	r* **GM**	A	*NW*	NH	52136	57136
150 137	r§ **GM**	A	*NW*	NH	52137	57137
150 138	r* **GM**	A	*NW*	NH	52138	57138
150 139	r* **GM**	A	*NW*	NH	52139	57139
150 140	r* **GM**	A	*NW*	NH	52140	57140
150 141	r* **GM**	A	*NW*	NH	52141	57141
150 142	r§ **GM**	A	*NW*	NH	52142	57142
150 143	r* **P**	A	*NW*	NH	52143	57143
150 144	r§ **P**	A	*NW*	NH	52144	57144
150 145	r§ **P**	A	*NW*	NH	52145	57145
150 146	r§ **RR**	A	*NW*	NH	52146	57146
150 147	r§ **P**	A	*NW*	NH	52147	57147
150 148	r§ **P**	A	*NW*	NH	52148	57148
150 149	r§ **P**	A	*NW*	NH	52149	57149
150 150	r§ **P**	A	*NW*	NH	52150	57150

CLASS 150/2   BREL SPRINTER

DMSL–DMS.

Engines: One Cummins NT855R5 of 210 kW (285 h.p.) per car.
Transmission: Hydraulic. Voith T211r with Gmeinder final drive.
Bogies: One BP38 and one BT38.
Gangways: Throughout.
Doors: Sliding.
Accommodation: 2+3 mainly unidirectional.
Dimensions: 20.06 x 2.82 m.
Maximum Speed: 75 m.p.h.

DMSL. Dia. DP242. Lot No. 31017 York 1986–87. –/73 1T (–70 1T*). 35.8 t.

DMS. Dia. DP243. Lot No. 31018 York 1986–7. –/76 (–/73§) and luggage space. 34.90 t.

g Fitted with global positioning system.

150 201	*	MT	A	NW	NH	52201	57201	
150 202		CE	A	CT	TS	52202	57202	
150 203	*	MT	A	NW	NH	52203	57203	
150 205	*	MT	A	NW	NH	52205	57205	
150 207	§	MT	A	NW	NH	52207	57207	
150 208		RR	P	SR	HA	52208	57208	
150 210		CE	A	CT	TS	52210	57210	
150 211	§	MT	A	NW	NH	52211	57211	
150 213		RR	P	AR	NC	52213	57213	Lord Nelson
150 214		CE	A	CT	TS	52214	57214	
150 215		GM	A	NW	NH	52215	57215	
150 216		CE	A	CT	TS	52216	57216	
150 217		RR	P	AR	NC	52217	57217	Oliver Cromwell
150 218	*§	GM	A	NW	NH	52218	57218	
150 219		RR	P	WW	CF	52219	57219	
150 220		CE	A	CT	TS	52220	57220	
150 221		RR	P	WW	CF	52221	57221	
150 223	*	GM	A	NW	NH	52223	57223	
150 224	*	GM	A	NW	NH	52224	57224	
150 225	§	GM	A	NW	NH	52225	57225	
150 227		RR	P	AR	NC	52227	57227	Sir Alf Ramsey
150 228		RR	P	SR	HA	52228	57228	
150 229		RR	P	AR	NC	52229	57229	George Borrow
150 230		RR	P	WW	CF	52230	57230	
150 231		RR	P	AR	NC	52231	57231	King Edmund
150 232		RR	P	WW	CF	52232	57232	
150 233		RR	P	WW	CF	52233	57233	
150 234		RR	P	WW	CF	52234	57234	
150 235		RR	P	AR	NC	52235	57235	Cardinal Wolsey
150 236		RR	P	WW	CF	52236	57236	
150 237		RR	P	AR	NC	52237	57237	Hereward the Wake
150 238		RR	P	WW	CF	52238	57238	
150 239		RR	P	WW	CF	52239	57239	
150 240		RR	P	WW	CF	52240	57240	
150 241		RR	P	WW	CF	52241	57241	
150 242		RR	P	WW	CF	52242	57242	
150 243		RR	P	WW	CF	52243	57243	
150 244		RR	P	WW	CF	52244	57244	
150 245		RR	P	SR	HA	52245	57245	
150 246		RR	P	WW	CF	52246	57246	
150 247		RR	P	WW	CF	52247	57247	
150 248		RR	P	WW	CF	52248	57248	
150 249		RR	P	WW	CF	52249	57249	
150 250		RR	P	SR	HA	52250	57250	
150 251		RR	P	WW	CF	52251	57251	
150 252		RR	P	SR	HA	52252	57252	
150 253		RR	P	WW	CF	52253	57253	

150 254		**RR**	P	*WW*	CF	52254	57254	
150 255		**RR**	P	*AR*	NC	52255	57255	Henry Blogg
150 256		**RR**	P	*SR*	HA	52256	57256	
150 257	§	**RR**	P	*AR*	NC	52257	57257	Queen Boadicea
150 258		**RR**	P	*SR*	HA	52258	57258	
150 259		**RR**	P	*SR*	HA	52259	57259	
150 260		**RR**	P	*SR*	HA	52260	57260	
150 261		**RR**	P	*WW*	CF	52261	57261	
150 262		**RR**	P	*SR*	HA	52262	57262	
150 263		**RR**	P	*WW*	CF	52263	57263	
150 264		**RR**	P	*SR*	HA	52264	57264	
150 265	g	**RR**	P	*WW*	CF	52265	57265	
150 266	g	**RR**	P	*CA*	CF	52266	57266	
150 267	g	**RR**	P	*CA*	CF	52267	57267	
150 268	g	**RR**	P	*CA*	CF	52268	57268	
150 269	g	**RR**	P	*CA*	CF	52269	57269	
150 270	g	**RR**	P	*CA*	CF	52270	57270	
150 271	g	**RR**	P	*CA*	CF	52271	57271	
150 272	g	**RR**	P	*CA*	CF	52272	57272	
150 273	g	**RR**	P	*CA*	CF	52273	57273	
150 274	g	**RR**	P	*CA*	CF	52274	57274	
150 275	g	**RR**	P	*CA*	CF	52275	57275	
150 276	g	**RR**	P	*CA*	CF	52276	57276	
150 277	g	**RR**	P	*CA*	CF	52277	57277	
150 278	g	**RR**	P	*CA*	CF	52278	57278	
150 279	g	**RR**	P	*CA*	CF	52279	57279	
150 280	g	**RR**	P	*CA*	CF	52280	57280	
150 281	g	**RR**	P	*CA*	CF	52281	57281	
150 282	g	**RR**	P	*CA*	CF	52282	57282	
150 283		**RR**	P	*SR*	HA	52283	57283	
150 284		**RR**	P	*SR*	HA	52284	57284	
150 285		**RR**	P	*SR*	HA	52285	57285	EDINBURGH–BATHGATE 1986–1996

CLASS 153   LEYLAND BUS SUPER SPRINTER

DMSL. Converted by Hunslet-Barclay, Kilmarnock from Class 155 two-car units.

Engines: One Cummins NT855R5 of 213 kW (285 h.p.) per car.
Transmission: Hydraulic. Voith T211r with Gmeinder final drive.
Bogies: One P3-10 and one BT38.
Gangways: Throughout.
Doors: Sliding plug.
Accommodation: 2+2 facing/unidirectional with wheelchair space.
Dimensions: 23.21 x 2.70 m.
Maximum Speed: 75 m.p.h.

52301–52335. DMSL. Dia. DX203. Lot No. 31026 1987–8. Converted under Lot No. 31115 1991–2. –/72 1TD (–/66 1TD†) + 3 tip-up seats. 41.2 t.
57301–57335. DMSL. Dia. DX203. Lot No. 31027 1987–8. Converted under Lot No. 31115 1991–2. –/72 1TD + 3 tip-up seats. 41.2 t.

Notes:

Cars numbered in the 573XX series have been renumbered by adding 50 to
the number so that the last two digits correspond with the set number.
Certain Central Trains units have been fitted with new-style seating and
certain Wales & West units have been fitted with Class 158-style seating.

153 301		**RR**	A	*NE*	HT	52301
153 302		**RR**	A	*WW*	CF	52302
153 303		**RR**	A	*WW*	CF	52303
153 304		**RR**	A	*NE*	HT	52304
153 305		**RR**	A	*WW*	CF	52305
153 306	†	**RR**	P	*AR*	NC	52306	Edith Cavell
153 307		**RR**	A	*NE*	HT	52307
153 308		**RR**	A	*WW*	CF	52308
153 309	†	**RR**	P	*AR*	NC	52309
153 310		**RR**	P	*NW*	NH	52310
153 311		**RR**	P	*AR*	NC	52311	John Constable
153 312		**RR**	A	*WW*	CF	52312
153 313		**RR**	P	*NW*	NH	52313
153 314	†	**RR**	P	*AR*	NC	52314	Delia Smith
153 315		**RR**	A	*NE*	HT	52315
153 316		**RR**	P	*NW*	NH	52316
153 317		**RR**	A	*NE*	HT	52317
153 318		**RR**	A	*WW*	CF	52318
153 319		**RR**	A	*NE*	HT	52319
153 320		**RR**	P	*CT*	TS	52320
153 321		**RR**	P	*CT*	TS	52321
153 322		**RR**	P	*AR*	NC	52322	Benjamin Britten
153 323		**RR**	P	*CT*	TS	52323
153 324		**RR**	P	*NW*	NH	52324
153 325		**RR**	P	*CT*	TS	52325
153 326	†	**RR**	P	*AR*	NC	52326	Ted Ellis
153 327		**RR**	A	*WW*	CF	52327
153 328		**RR**	A	*NE*	HT	52328
153 329		**RR**	P	*CT*	TS	52329
153 330		**RR**	P	*NW*	NH	52330
153 331		**RR**	A	*NE*	HT	52331
153 332		**RR**	P	*NW*	NH	52332
153 333		**RR**	P	*CT*	TS	52333
153 334		**RR**	P	*CT*	TS	52334
153 335	†	**RR**	P	*AR*	NC	52335	MICHAEL PALIN
153 351		**RR**	A	*NE*	HT	57351
153 352		**RR**	A	*NE*	HT	57352
153 353		**RR**	A	*WW*	CF	57353
153 354		**RR**	P	*CT*	TS	57354
153 355		**RR**	A	*WW*	CF	57355
153 356		**RR**	P	*CT*	TS	57356
153 357		**RR**	A	*NE*	HT	57357
153 358		**RR**	P	*NW*	NH	57358
153 359		**RR**	P	*NW*	NH	57359
153 360		**RR**	P	*NW*	NH	57360

153 361	**RR**	P	*NW*	NH	57361
153 362	**RR**	A	*WW*	CF	57362
153 363	**RR**	P	*NW*	NH	57363
153 364	**RR**	P	*CT*	TS	57364
153 365	**RR**	P	*CT*	TS	57365
153 366	**RR**	P	*CT*	TS	57366
153 367	**RR**	P	*NW*	NH	57367
153 368	**RR**	A	*WW*	CF	57368
153 369	**RR**	P	*CT*	TS	57369
153 370	**RR**	A	*WW*	CF	57370
153 371	**RR**	P	*CT*	TS	57371
153 372	**RR**	A	*WW*	CF	57372
153 373	**RR**	A	*WW*	CF	57373
153 374	**RR**	A	*WW*	CF	57374
153 375	**RR**	P	*CT*	TS	57375
153 376	**RR**	P	*CT*	TS	57376
153 377	**RR**	A	*WW*	CF	57377
153 378	**RR**	A	*NE*	HT	57378
153 379	**RR**	P	*CT*	TS	57379
153 380	**RR**	A	*WW*	CF	57380
153 381	**RR**	P	*CT*	TS	57381
153 382	**RR**	A	*WW*	CF	57382
153 383	**RR**	P	*CT*	TS	57383
153 384	**RR**	P	*CT*	TS	57384
153 385	**RR**	P	*CT*	TS	57385

CLASS 155   LEYLAND BUS SUPER SPRINTER

DMSL–DMS.

Engines: One Cummins NT855R5 of 213 kW (285 h.p.) per car.
Transmission: Hydraulic. Voith T211r with Gmeinder final drive.
Bogies: One P3-10 and one BT38.
Gangways: Throughout.
Doors: Sliding plug.
Accommodation: 2+2 facing/unidirectional with wheelchair space in DMSL.
Dimensions: 23.21 x 2.70 m.
Maximum Speed: 75 m.p.h.

DMSL. Dia. DP248. Lot No. 31057 1988. –/80 1TD. 39.0 t.
DMS. Dia. DP249. Lot No. 31058 1988. –/80 and parcels area. 38.7 t.

Note: These units are owned by West Yorkshire PTE, although managed by Porterbrook Leasing Company.

155 341	**Y**	P	*NE*	NL	52341	57341
155 342	**Y**	P	*NE*	NL	52342	57342
155 343	**Y**	P	*NE*	NL	52343	57343
155 344	**Y**	P	*NE*	NL	52344	57344
155 345	**Y**	P	*NE*	NL	52345	57345
155 346	**Y**	P	*NE*	NL	52346	57346
155 347	**Y**	P	*NE*	NL	52347	57347

CLASS 156 METRO-CAMMELL SUPER SPRINTER

DMSL–DMS.

Engines: One Cummins NT855R5 of 210 kW (285 h.p.) per car.
Transmission: Hydraulic. Voith T211r with Gmeinder final drive.
Bogies: One P3-10 and one BT38.
Gangways: Throughout.
Doors: Sliding.
Accommodation: 2+2 facing/unidirectional with wheelchair space in DMSL.
Dimensions: 23.03 x 2.73 m.
Maximum Speed: 75 m.p.h.

DMSL. Dia. DP244. Lot No. 31028 1988–9. –/74 (–/72 †, –/70 § •) 1TD. 36.1 t.
DMS. Dia. DP245. Lot No. 31029 1987–9. 35.5 t. –/76 (72 •) + parcels area.

Notes: 156 500–514 are owned by Strathclyde PTE, although managed by
Angel Trains Contracts. Units reliveried in **RE** or **RN** livery have been fitted with
new-style seats.

156 401	†	**RE**	P	*CT*	TS	52401 57401
156 402	†	**RE**	P	*CT*	TS	52402 57402
156 403	†	**RE**	P	*CT*	TS	52403 57403
156 404	†	**RE**	P	*CT*	TS	52404 57404
156 405	†	**RE**	P	*CT*	TS	52405 57405
156 406	†	**RE**	P	*CT*	TS	52406 57406
156 407	†	**RE**	P	*CT*	TS	52407 57407
156 408	†	**RE**	P	*CT*	TS	52408 57408
156 409	†	**RE**	P	*CT*	TS	52409 57409
156 410	†	**RE**	P	*CT*	TS	52410 57410
156 411	†	**RE**	P	*CT*	TS	52411 57411
156 412	†	**RE**	P	*CT*	TS	52412 57412
156 413	†	**RE**	P	*CT*	TS	52413 57413
156 414	†	**RE**	P	*CT*	TS	52414 57414
156 415	†	**RE**	P	*CT*	TS	52415 57415
156 416	†	**RE**	P	*CT*	TS	52416 57416
156 417	†	**RE**	P	*CT*	TS	52417 57417
156 418	†	**RE**	P	*CT*	TS	52418 57418
156 419	†	**RE**	P	*CT*	TS	52419 57419
156 420	§	**RN**	P	*NW*	NH	52420 57420
156 421	§	**RN**	P	*NW*	NH	52421 57421
156 422	†	**RE**	P	*CT*	TS	52422 57422
156 423	§	**RN**	P	*NW*	NH	52423 57423
156 424	§	**RN**	P	*NW*	NH	52424 57424
156 425	§	**RN**	P	*NW*	NH	52425 57425
156 426	§	**RN**	P	*NW*	NH	52426 57426
156 427	§	**RN**	P	*NW*	NH	52427 57427
156 428	§	**RN**	P	*NW*	NH	52428 57428
156 429	§	**RN**	P	*NW*	NH	52429 57429
156 430		**P**	A	*SR*	CK	52430 57430
156 431	r	**P**	A	*SR*	CK	52431 57431
156 432	r	**P**	A	*SR*	CK	52432 57432

156 433		**CC**	A	*SR*	CK	52433	57433	The Kilmarnock Edition
156 434	r	**P**	A	*SR*	CK	52434	57434
156 435	r	**P**	A	*SR*	CK	52435	57435
156 436	r	**P**	A	*SR*	CK	52436	57436
156 437		**P**	A	*SR*	CK	52437	57437
156 438		**P**	A	*NE*	NL	52438	57438
156 439		**P**	A	*SR*	CK	52439	57439
156 440	§	**RN**	P	*NW*	NH	52440	57440
156 441	§	**RN**	P	*NW*	NH	52441	57441
156 442		**P**	A	*SR*	CK	52442	57442
156 443		**P**	A	*NE*	HT	52443	57443
156 444		**P**	A	*NE*	HT	52444	57444
156 445	r	**P**	A	*SR*	CK	52445	57445
156 446	r	**P**	A	*SR*	IS	52446	57446
156 447	r	**P**	A	*SR*	HA	52447	57447
156 448		**P**	A	*NE*	CK	52448	57448
156 449	r	**P**	A	*SR*	CK	52449	57449
156 450	r	**P**	A	*SR*	CK	52450	57450
156 451		**P**	A	*NE*	HT	52451	57451
156 452	§	**RN**	P	*NW*	NH	52452	57452
156 453	r	**P**	A	*SR*	CK	52453	57453
156 454		**P**	A	*NE*	HT	52454	57454
156 455	†	**RN**	P	*NW*	NH	52455	57455
156 456	r	**P**	A	*SR*	CK	52456	57456
156 457	r	**P**	A	*SR*	IS	52457	57457
156 458	r	**P**	A	*SR*	IS	52458	57458
156 459	†	**RN**	P	*NW*	NH	52459	57459
156 460	†	**RN**	P	*NW*	NH	52460	57460
156 461	†	**RN**	P	*NW*	NH	52461	57461
156 462	r	**P**	A	*SR*	CK	52462	57462
156 463		**P**	A	*NE*	HT	52463	57463
156 464	§	**RN**	P	*NW*	NH	52464	57464
156 465	r	**P**	A	*SR*	CK	52465	57465	Bonnie Prince Charlie
156 466	§	**RN**	P	*NW*	NH	52466	57466
156 467	r	**P**	A	*SR*	CK	52467	57467
156 468		**P**	A	*NE*	NL	52468	57468
156 469		**P**	A	*NE*	HT	52469	57469
156 470		**P**	A	*NE*	NL	52470	57470
156 471		**P**	A	*NE*	NL	52471	57471
156 472		**P**	A	*NE*	NL	52472	57472
156 473		**P**	A	*NE*	NL	52473	57473
156 474	r•	**P**	A	*SR*	IS	52474	57474
156 475		**P**	A	*NE*	NL	52475	57475
156 476		**P**	A	*SR*	CK	52476	57476
156 477	r•	**P**	A	*SR*	IS	52477	57477	HIGHLAND FESTIVAL
156 478	r•	**P**	A	*SR*	IS	52478	57478
156 479		**P**	A	*NE*	NL	52479	57479
156 480		**P**	A	*NE*	NL	52480	57480
156 481		**P**	A	*NE*	NL	52481	57481
156 482		**P**	A	*NE*	NL	52482	57482
156 483		**P**	A	*NE*	NL	52483	57483

156 484		P	A	*NE*	NL	52484 57484
156 485	r•	P	A	*SR*	CK	52485 57485
156 486	r•	P	A	*NE*	NL	52486 57486
156 487		P	A	*NE*	NL	52487 57487
156 488		P	A	*NE*	NL	52488 57488
156 489		P	A	*NE*	NL	52489 57489
156 490		P	A	*NE*	NL	52490 57490
156 491		P	A	*NE*	NL	52491 57491
156 492	r•	P	A	*SR*	CK	52492 57492
156 493	r•	P	A	*SR*	CK	52493 57493
156 494	r•	P	A	*SR*	CK	52494 57494
156 495	r•	P	A	*SR*	CK	52495 57495
156 496	r•	P	A	*SR*	CK	52496 57496
156 497		P	A	*NE*	NL	52497 57497
156 498		P	A	*NE*	NL	52498 57498
156 499	r•	P	A	*SR*	IS	52499 57499
156 500	r•	P	A	*SR*	CK	52500 57500
156 501		S	A	*SR*	CK	52501 57501
156 502		S	A	*SR*	CK	52502 57502
156 503		S	A	*SR*	CK	52503 57503
156 504	r•	S	A	*SR*	CK	52504 57504
156 505	r•	S	A	*SR*	CK	52505 57505
156 506		S	A	*SR*	CK	52506 57506
156 507		S	A	*SR*	CK	52507 57507
156 508		S	A	*SR*	CK	52508 57508
156 509		S	A	*SR*	CK	52509 57509
156 510		S	A	*SR*	CK	52510 57510
156 511		S	A	*SR*	CK	52511 57511
156 512		S	A	*SR*	CK	52512 57512
156 513		S	A	*SR*	CK	52513 57513
156 514		S	A	*SR*	CK	52514 57514

CLASS 158/0   BREL EXPRESS

DMSL (B)–DMSL (A) or DMCL–DMSL*§ or DMSL (B)–MSL–DMSL (A).

Engines: One Cummins NTA855R of 260 kW (350 h.p.) or 300 kW (400 h.p.)
• One Perkins 2006-TWH of 260 kW (350 h.p.)†) per car.
Transmission: Hydraulic. Voith T211r with Gmeinder final drive.
Bogies: One BREL P4 and one BREL T4 per car.
Gangways: Throughout.
Doors: Sliding plug.
Accommodation: 2+2 facing/unidirectional (first & standard classes).
Dimensions: 23.21 x 2.70 m.
Maximum Speed: 90 m.p.h.

DMSL (B).. Dia. DP252. Lot No. 31051 Derby 1990–2. –/68 + wheelchair space
1TD. Public telephone and trolley space. 38.5 t.
DMCL.. Dia. DP252. Lot No. 31051 Derby 1989–90. 15/51*, 9/51§ + wheelchair space 1TD. Public telephone and trolley space. 38.5 t.
MSL. Dia. DR207. Lot No. 31050 Derby 1991. 38 t. –/70 2T.

DMSL (A). Dia. DP251. Lot No. 31052 Derby 1990–92. –/70 1T and parcels area. 37.8 t.

158 701	*	**RE**	P	*SR*	HA	52701	57701 The Scottish Claymores
158 702	*	**RE**	P	*SR*	HA	52702	57702
158 703	*	**RE**	P	*SR*	HA	52703	57703
158 704	*	**RE**	P	*SR*	HA	52704	57704
158 705	*	**RE**	P	*SR*	HA	52705	57705
158 706	*	**RE**	P	*SR*	HA	52706	57706
158 707	*	**RE**	P	*SR*	HA	52707	57707
158 708	*	**RE**	P	*SR*	HA	52708	57708
158 709	*	**RE**	P	*SR*	HA	52709	57709
158 710	*	**RE**	P	*SR*	HA	52710	57710
158 711	*	**RE**	P	*SR*	HA	52711	57711
158 712	*	**RE**	P	*SR*	HA	52712	57712
158 713	*	**RE**	P	*SR*	HA	52713	57713
158 714	*	**RE**	P	*SR*	HA	52714	57714
158 715	*	**RE**	P	*SR*	HA	52715	57715 Haymarket
158 716	*	**RE**	P	*SR*	HA	52716	57716
158 717	*	**RE**	P	*SR*	HA	52717	57717
158 718	*	**RE**	P	*SR*	HA	52718	57718
158 719	*	**RE**	P	*SR*	HA	52719	57719
158 720	*	**RE**	P	*SR*	HA	52720	57720
158 721	*	**RE**	P	*SR*	HA	52721	57721
158 722	*	**RE**	P	*SR*	HA	52722	57722
158 723	*	**RE**	P	*SR*	HA	52723	57723
158 724	*	**RE**	P	*SR*	HA	52724	57724
158 725	*	**RE**	P	*SR*	HA	52725	57725
158 726	*	**RE**	P	*SR*	HA	52726	57726
158 727	*	**RE**	P	*SR*	HA	52727	57727
158 728	*	**RE**	P	*SR*	HA	52728	57728
158 729	*	**RE**	P	*SR*	HA	52729	57729
158 730	*	**RE**	P	*SR*	HA	52730	57730
158 731	*	**RE**	P	*SR*	HA	52731	57731
158 732	*	**RE**	P	*SR*	HA	52732	57732
158 733	*	**RE**	P	*SR*	HA	52733	57733
158 734	*	**RE**	P	*SR*	HA	52734	57734
158 735	*	**RE**	P	*SR*	HA	52735	57735
158 736	*	**RE**	P	*SR*	HA	52736	57736
158 737	*	**RE**	P	*SR*	HA	52737	57737
158 738	*	**RE**	P	*SR*	HA	52738	57738
158 739	*	**RE**	P	*SR*	HA	52739	57739
158 740	*	**RE**	P	*SR*	HA	52740	57740
158 741	*	**RE**	P	*SR*	HA	52741	57741
158 742	*	**RE**	P	*SR*	HA	52742	57742
158 743	*	**RE**	P	*SR*	HA	52743	57743
158 744	*	**RE**	P	*SR*	HA	52744	57744
158 745	*	**RE**	P	*SR*	HA	52745	57745
158 746	*	**RE**	P	*SR*	HA	52746	57746
158 747	§	**RE**	P	*VX*	NH	52747	57747

158 748	§	**RE**	P	*VX*	NH	52748		57748
158 749	§	**RE**	P	*VX*	NH	52749		57749
158 750	§	**RE**	P	*VX*	NH	52750		57750
158 751	§	**RE**	P	*VX*	NH	52751		57751
158 752		**RE**	P	*NW*	NH	52752		57752
158 753		**RE**	P	*NW*	NH	52753		57753
158 754		**RE**	P	*NW*	NH	52754		57754
158 755		**RE**	P	*NW*	NH	52755		57755
158 756		**RE**	P	*NW*	NH	52756		57756
158 757		**RE**	P	*NW*	NH	52757		57757
158 758		**RE**	P	*NW*	NH	52758		57758
158 759		**RE**	P	*NW*	NH	52759		57759
158 760		**RE**	P	*NE*	NL	52760		57760
158 761		**RE**	P	*NE*	NL	52761		57761
158 762		**RE**	P	*NE*	NL	52762		57762
158 763		**RE**	P	*NE*	NL	52763		57763
158 764		**RE**	P	*NE*	NL	52764		57764
158 765		**RE**	P	*NE*	NL	52765		57765
158 766		**RE**	P	*NE*	NL	52766		57766
158 767		**RE**	P	*NE*	NL	52767		57767
158 768		**RE**	P	*NE*	NL	52768		57768
158 769		**RE**	P	*NE*	NL	52769		57769
158 770		**RE**	P	*NE*	NL	52770		57770
158 771		**RE**	P	*NE*	HT	52771		57771
158 772		**RE**	P	*NE*	NL	52772		57772
158 773		**RE**	P	*NE*	NL	52773		57773
158 774		**RE**	P	*NE*	HT	52774		57774
158 775		**RE**	P	*NE*	HT	52775		57775
158 776		**RE**	P	*NE*	HT	52776		57776
158 777		**RE**	P	*NE*	HT	52777		57777
158 778		**RE**	P	*NE*	HT	52778		57778
158 779		**RE**	P	*NE*	HT	52779		57779
158 780	r	**RE**	A	*CT*	NC	52780		57780
158 781	r	**RE**	P	*NE*	HT	52781		57781
158 782	r	**RE**	A	*CT*	NC	52782		57782
158 783	r	**RE**	A	*CT*	NC	52783		57783
158 784	r	**RE**	A	*CT*	NC	52784		57784
158 785	r	**RE**	A	*CT*	NC	52785		57785
158 786	r	**RE**	A	*CT*	NC	52786		57786
158 787	r	**RE**	A	*CT*	NC	52787		57787
158 788	r	**RE**	A	*CT*	NC	52788		57788
158 789	r	**RE**	A	*CT*	NC	52789		57789
158 790	r	**RE**	A	*CT*	NC	52790		57790
158 791	r	**RE**	A	*CT*	NC	52791		57791
158 792	r	**RE**	A	*CT*	NC	52792		57792
158 793	r	**RE**	A	*CT*	NC	52793		57793
158 794	r	**RE**	A	*CT*	NC	52794		57794
158 795	r	**RE**	A	*CT*	NC	52795		57795
158 796	r	**RE**	A	*CT*	NC	52796		57796
158 797	r	**RE**	A	*CT*	NC	52797		57797
158 798		**RE**	P	*NE*	HT	52798	58715	57798

PLATFORM 5 PUBLISHING LIMITED
MAIL ORDER LIST

NEW TITLES

BR Pocket Book No.1: Locomotives	£2.60
BR Pocket Book No.2: Coaching Stock	£2.60
BR Pocket Book No.3: DMUs & Light Rail Systems	£2.60
BR Pocket Book No.4: Electric Multiple Units	£2.60
British Railways Locomotives & Coaching Stock 1998 **MARCH**	£10.50
Light Rail Review 8 **MARCH**	£9.50
British Railways Locomotives - The First 12 Years (SCTP)	£18.95
London Underground Rolling Stock (Capital)	£9.95
Railways around Lake Luzern (Bairstow)	£9.95
London Tilbury & Southend Railway Part 2 (Kay)	£9.95
Johnson's Atlas & Gazetteer of the Railways of Ireland (Midland)	£19.99
The Londonderry & Lough Swilly Railway (Midland)	£8.99
The Cavan & Leitrim Railway (Midland)	£8.99
The 1998 Cowie Bus Handbook (British Bus)	£15.00
The Fire Brigade Handbook: Special Appliances Vol. 2 (British Bus)	£12.50

MODERN BRITISH RAILWAY TITLES

Preserved Locomotives of British Railways 9th ed.	£7.95
Preserved Coaching Stock Part 1: BR Design Stock	£7.95
Preserved Coaching Stock Part 2: Pre-Nationalisation Stock	£8.95
Diesel & Electric Loco Register 3rd edition	£7.95
Valley Lines - The People's Railway	£9.95
Air Braked Series Wagon Fleet (SCTP)	£7.95
Departmental Coaching Stock 5th edition (SCTP)	£6.95
On-Track Plant on British Railways 5th edition (SCTP)	£7.95
Engineers Series Wagon Fleet 970000-999999 (SCTP)	£6.95
British Rail Wagon Fleet - B-Prefix Series (SCTP)	£6.95
British Rail Internal Users (SCTP)	£7.95
Private Owner Wagons Volume 1 (Metro)	£7.95
Miles & Chains Volume 2 - London Midland (Milepost)	£1.95
Miles & Chains Volume 3 - Scottish (Milepost)	£1.95
Miles & Chains Volume 5 - Southern (Milepost)	£1.95

OVERSEAS RAILWAYS

High Speed in Europe	£9.95
High Speed in Japan	£16.95

European Handbook No. 1: Benelux Railways 3rd edition £10.50
European Handbook No. 3: Austrian Railways 3rd edition £10.50
European Handbook No. 5: Swiss Railways 2nd edition £13.50
European Handbook No. 6: Italian Railways 1st edition £13.50
European Handbook No. 7: Irish Railways 1st edition £9.95
The Railways of Greece (Simms) £8.10
The Railways of Corsica (Simms) £5.10
Railways in the Austrian Tirol (Bairstow) £8.95
Irish Railways In Colour: From Steam to Diesel 1955-1967 (Midland) £16.99
Irish Railways In Colour: A Second Glance 1947-1970 (Midland) £19.99
Locomotives & Railcars of Bord Na Mona (Midland) £4.99

METRO SYSTEMS
World Metro Systems 2nd ed. (Capital) £10.95
The Twopenny Tube (Capital) .... [History of the Central Line] £5.95
Circles Under the Clyde (Capital) .... [Glasgow Subway] £15.95
Underground Official Handbook (Capital) £7.95
Docklands Light Rail Official Handbook (Capital) £7.95
Paris Metro Handbook (Capital) £7.95
The Berlin S-Bahn (Capital) £7.50
The Berlin U-Bahn (Capital) £7.50
Underground Architecture (Capital) £25.00
Mr Beck's Underground Map (Capital) £10.95

LIGHT RAIL TRANSIT AND TRAMS
Tram to Supertram .... [Sheffield Trams] £4.95
Light Rail Review 3 .... £7.50
Light Rail Review 4 .... £7.50
Light Rail Review 5 .... £7.50
Light Rail Review 6 .... £7.50
Light Rail Review 7 .... £8.95
Manx Electric £8.95
Light Rail in Europe (Capital) £9.95
London Tramways (Capital) £19.95
Tramway & Light Railway Atlas Germany 1996 (Blickpunkt Strassenbahn/LRTA) £10.45
The Tramways of Portugal (LRTA) £9.05

ATLASES, MAPS AND TRACK DIAGRAMS
Railway Track Diagrams No. 1: Scotland & Isle of Man (Quail) £6.50
British Railway Track Diagrams No. 4: Midland - 1990 Reprint (Quail) £6.95
Railway Track Diagrams No. 6: Ireland (Quail) £5.50

London Transport Railway Track Map (Quail) £1.75
Czech Republic & Slovakia Railway Map (Quail) £1.70
Berlin Track Map (Quail) £2.20
Moscow Railway Map (Quail) £2.20
Portugal Railway Map (Quail) £2.00
Greece Railway Map (Quail) £1.70
Poland Railway Map (Quail) £2.00
European Railway Atlas: France, Benelux (Ian Allan) £10.99
Track Diagram - South Yorkshire Supertram (HRT Rail Sales) £1.50
Track Diagram - Blackpool & Fleetwood (HRT Rail Sales) £1.00
Track Diagram - Tyne & Wear (HRT Rail Sales) £2.00

HISTORICAL RAILWAY TITLES
Steam Days on BR 1 - The Midland Line in Sheffield £4.95
Rails along the Sea Wall .... [Dawlish-Teignmouth Pictorial] £4.95
The Rolling Rivers £6.95
British Baltic Tanks £6.95
London Tilbury & Southend Railway Part 1 (Kay) £9.95
Signalling Atlas and Signal Box Directory Great Britain & Ireland (Kay) £9.95
Midland Railway System Maps: Leicester-London (Kay) £8.95
Mechanical Railway Signalling Part 1 (Kay) £9.50
Rails in the Isle of Wight (Midland) £16.99

ROAD TRANSPORT
London Coach Handbook (Capital) £15.00
London's Utility Buses (Capital) £19.95
London's Wartime Gas Buses (Capital) £5.95
Truckin' Round Scotland (Arthur Southern) £10.95
Bus Review 12 (Bus Enthusiast) £7.50
The Yorkshire Bus Handbook (British Bus) £12.50
The Ireland & Islands Bus Handbook (British Bus) £9.95
The Fire Brigade Handbook: Special Appliances Vol. 1 (British Bus) £12.50

RAMBLING
Rambles by Rail 2 - Liskeard-Looe £1.95
Rambles by Rail 4 - The New Forest £1.95
Buxton Spa Line Rail Rambles £1.20

POSTCARDS
Sheffield Supertram - Car No. 12 crosses a bridge over Sheffield Canal £0.30
Manchester Metrolink - Car No. 1021 in Aytoun Street £0.30

Quantity	Title		Price	Total
		SUB-TOTAL		
	Postage & Packing (see below for details)			
		TOTAL REMITTANCE		

Name: ....

Address: ....

....

.... Postcode: ....

Telephone No.: .... (Home) .... (Work)

Payment (Delete as appropriate)

I enclose my cheque (drawn on a UK bank) / British postal order for £ .... payable to:

'PLATFORM 5 PUBLISHING LTD'

Please debit my Visa / MasterCard / Access / Delta / Eurocard credit card for £ ....

Card No: .... Card Expiry Date: ....

Signature: .... Date: ....

Minimum credit card order accepted - £3.00.

Please send your remittance to:

Platform 5 Mail Order Department (PB)
3 Wyvern House, Sark Road
SHEFFIELD, S2 4HG, ENGLAND

If paying by credit card we can accept payment by post, or by telephone: UK - 0114 255 2625, Overseas - +44 114 255 2625, or fax: UK - 0114 255 2471, Overseas - +44 114 255 2471.

Postage & packing: please add: 10% UK (2nd Class); 20% Europe (Airmail); 30% Rest of World (Airfreight); 50% Rest of World (Airmail). If p&p works out at less than 40p, then please send 40p, this is the minimum post & packing accepted.

Please note that we cannot accept foreign currency cheques.

NOTE. When ordering publications in conjunction with a **Today's Railways** subscription offer please add on post & packing **before** deducting the voucher. Vouchers may **not** be combined.

Details correct as at 31st January 1998. Prices are not guaranteed and we reserve the right to alter details without further notification. Please allow 28 days for delivery in the UK

158 799		**RE**	P	*NE*	HT	52799	58716	57799
158 800		**RE**	P	*NE*	HT	52800	58717	57800
158 801		**RE**	P	*NE*	HT	52801	58701	57801
158 802		**RE**	P	*NE*	HT	52802	58702	57802
158 803		**RE**	P	*NE*	HT	52803	58703	57803
158 804		**RE**	P	*NE*	HT	52804	58704	57804
158 805		**RE**	P	*NE*	HT	52805	58705	57805
158 806		**RE**	P	*NE*	HT	52806	58706	57806
158 807		**RE**	P	*NE*	HT	52807	58707	57807
158 808		**RE**	P	*NE*	HT	52808	58708	57808
158 809		**RE**	P	*NE*	HT	52809	58709	57809
158 810		**RE**	P	*NE*	HT	52810	58710	57810
158 811		**RE**	P	*NE*	HT	52811	58711	57811
158 812		**RE**	P	*NE*	HT	52812	58712	57812
158 813		**RE**	P	*NE*	HT	52813	58713	57813
158 814		**RE**	P	*NE*	HT	52814	58714	57814
158 815	†	**RE**	A	*WW*	CF	52815		57815
158 816	†	**RE**	A	*WW*	CF	52816		57816
158 817	†	**RE**	A	*WW*	CF	52817		57817
158 818	†	**RE**	A	*WW*	CF	52818		57818
158 819	†	**RE**	A	*WW*	CF	52819		57819
158 820	†	**RE**	A	*WW*	CF	52820		57820
158 821	†	**RE**	A	*WW*	CF	52821		57821
158 822	†	**RE**	A	*WW*	CF	52822		57822
158 823	†	**RE**	A	*WW*	CF	52823		57823
158 824	†	**RE**	A	*WW*	CF	52824		57824
158 825	†	**RE**	A	*WW*	CF	52825		57825
158 826	†	**RE**	A	*WW*	CF	52826		57826
158 827	†	**RE**	A	*WW*	CF	52827		57827
158 828	†	**RE**	A	*WW*	CF	52828		57828
158 829	†	**RE**	A	*WW*	CF	52829		57829
158 830	†	**RE**	A	*WW*	CF	52830		57830
158 831	†	**RE**	A	*WW*	CF	52831		57831
158 832	†	**RE**	A	*WW*	CF	52832		57832
158 833	†	**RE**	A	*WW*	CF	52833		57833
158 834	†	**RE**	A	*WW*	CF	52834		57834
158 835	†	**RE**	A	*WW*	CF	52835		57835
158 836	†	**RE**	A	*WW*	CF	52836		57836
158 837	†	**RE**	A	*WW*	CF	52837		57837
158 838	†	**RE**	A	*WW*	CF	52838		57838
158 839	†	**RE**	A	*WW*	CF	52839		57839
158 840	†	**RE**	A	*WW*	CF	52840		57840
158 841	†	**RE**	A	*WW*	CF	52841		57841
158 842	†r	**RE**	A	*WW*	CF	52842		57842
158 843	†r	**RE**	A	*WW*	CF	52843		57843
158 844	†r	**RE**	A	*CT*	NC	52844		57844
158 845	†r	**RE**	A	*CT*	NC	52845		57845
158 846	†r	**RE**	A	*CT*	NC	52846		57846
158 847	†r	**RE**	A	*CT*	NC	52847		57847
158 848	†r	**RE**	A	*CT*	NC	52848		57848
158 849	†r	**RE**	A	*CT*	NC	52849		57849

158 850	†r	**RE**	A	*CT*	NC	52850	57850
158 851	†r	**RE**	A	*CT*	NC	52851	57851
158 852	†r	**RE**	A	*CT*	NC	52852	57852
158 853	†r	**RE**	A	*CT*	NC	52853	57853
158 854	†r	**RE**	A	*CT*	NC	52854	57854
158 855	†r	**RE**	A	*CT*	NC	52855	57855
158 856	†r	**RE**	A	*CT*	NC	52856	57856
158 857	†r	**RE**	A	*CT*	NC	52857	57857
158 858	†r	**RE**	A	*CT*	NC	52858	57858
158 859	†r	**RE**	A	*CT*	NC	52859	57859
158 860	†r	**RE**	A	*CT*	NC	52860	57860
158 861	†r	**RE**	A	*CT*	NC	52861	57861
158 862	†r	**RE**	A	*CT*	NC	52862	57862
158 863	•	**RE**	A	*WW*	CF	52863	57863
158 864	•	**RE**	A	*WW*	CF	52864	57864
158 865	•	**RE**	A	*WW*	CF	52865	57865
158 866	•	**RE**	A	*WW*	CF	52866	57866
158 867	•	**RE**	A	*WW*	CF	52867	57867
158 868	•	**RE**	A	*WW*	CF	52868	57868
158 869	•	**RE**	A	*WW*	CF	52869	57869
158 870	•	**RE**	A	*WW*	CF	52870	57870
158 871	•	**RE**	A	*WW*	CF	52871	57871
158 872	•	**RE**	A	*WW*	CF	52872	57872

CLASS 158/9   BREL EXPRESS

DMSL–DMS. Units leased by West Yorkshire PTE. Details as for Class 158/0
except for seating layout and toilets.

DMSL.. Dia. DP252. Lot No. 31051 Derby 1990–2. –/70 + wheelchair space
1TD. Public telephone and trolley space. 38.1 t.
DMS. Dia. DP251. Lot No. 31052 Derby 1990–92. –/72 and parcels area. 37.8 t.

Note: Although these units are leased by West Yorkshire PTE, they are
managed by Porterbrook Leasing Company.

158 901	**Y**	P	*NE*	NL	52901	57901
158 902	**Y**	P	*NE*	NL	52902	57902
158 903	**Y**	P	*NE*	NL	52903	57903
158 904	**Y**	P	*NE*	NL	52904	57904
158 905	**Y**	P	*NE*	NL	52905	57905
158 906	**Y**	P	*NE*	NL	52906	57906
158 907	**Y**	P	*NE*	NL	52907	57907
158 908	**Y**	P	*NE*	NL	52908	57908
158 909	**Y**	P	*NE*	NL	52909	57909
158 910	**Y**	P	*NE*	NL	52910	57910

CLASS 159   BREL EXPRESS

DMCL–MSL–DMSL. Built as Class 158 by BREL. Converted before entering
passenger service to Class 159 by Rosyth Dockyard.

Engines: One Cummins NTA855R of 300 kW (400 h.p.) per car.
Transmission: Hydraulic. Voith T211r with Gmeinder final drive.
Bogies: One BREL P4 and one BREL T4 per car.
Gangways: Throughout.
Doors: Sliding plug.
Accommodation: 2+2 facing/unidirectional (standard class), 2+1 facing (first class).
Dimensions: 23.21 x 2.82 m.
Maximum Speed: 90 m.p.h.

DMCL.. Dia. DP322. Lot No. 31051 Derby 1992. 24/28 1TD. 38.5 t.
MSL. Dia. DR209. Lot No. 31050 Derby 1992. 38 t. –/72 2T.
DMSL. Dia. DP260. Lot No. 31052 Derby 1992. –/72 1T and parcels area. 37.8 t.

159 001	**NW**	P	*SW*	SA	52873 58718 57873	CITY OF EXETER
159 002	**NW**	P	*SW*	SA	52874 58719 57874	CITY OF SALISBURY
159 003	**NW**	P	*SW*	SA	52875 58720 57875	TEMPLECOMBE
159 004	**NW**	P	*SW*	SA	52876 58721 57876	BASINGSTOKE AND DEANE
159 005	**NW**	P	*SW*	SA	52877 58722 57877	
159 006	**NW**	P	*SW*	SA	52878 58723 57878	
159 007	**NW**	P	*SW*	SA	52879 58724 57879	
159 008	**NW**	P	*SW*	SA	52880 58725 57880	
159 009	**NW**	P	*SW*	SA	52881 58726 57881	
159 010	**NW**	P	*SW*	SA	52882 58727 57882	
159 011	**NW**	P	*SW*	SA	52883 58728 57883	
159 012	**NW**	P	*SW*	SA	52884 58729 57884	
159 013	**NW**	P	*SW*	SA	52885 58730 57885	
159 014	**NW**	P	*SW*	SA	52886 58731 57886	
159 015	**NW**	P	*SW*	SA	52887 58732 57887	
159 016	**NW**	P	*SW*	SA	52888 58733 57888	
159 017	**NW**	P	*SW*	SA	52889 58734 57889	
159 018	**NW**	P	*SW*	SA	52890 58735 57890	
159 019	**NW**	P	*SW*	SA	52891 58736 57891	
159 020	**NW**	P	*SW*	SA	52892 58737 57892	
159 021	**NW**	P	*SW*	SA	52893 58738 57893	
159 022	**NW**	P	*SW*	SA	52894 58739 57894	

CLASS 165/0   BREL NETWORK TURBO

DMCL–DMS or DMCL–MS–DMS. Built for Chiltern Line services.

Engines: One Perkins 2006-TWH of 260 kW (350 h.p.) per car.
Transmission: Hydraulic. Voith T211r with Gmeinder final drive.
Bogies: One BREL P3 and one BREL T3 per car.
Gangways: Within unit only.
Doors: Sliding plug.
Accommodation: 2+3 facing/unidirectional (standard class), 2+2 facing (first class).
Dimensions: 23.50 x 2.85 m.
Maximum Speed: 75 m.p.h.

58801–58822. 58873–58878. DMCL. Dia. DP319. Lot No. 31087 York 1990. 16/72 1T. 37.0 t.
58823–58833. DMCL. Dia. DP320. Lot No. 31089 York 1991–1992. 24/60 1T. 37.0 t.
MS. Dia. DR208. Lot No. 31090 York 1991–1992. 106S. 37.0 t.
DMS. Dia. DP253. Lot No. 31088 York 1991–1992. 98S. 37.0 t.

165 001	**NW**	A	*TT*	RG	58801		58834
165 002	**NW**	A	*TT*	RG	58802		58835
165 003	**NW**	A	*TT*	RG	58803		58836
165 004	**NW**	A	*TT*	RG	58804		58837
165 005	**NW**	A	*TT*	RG	58805		58838
165 006	**NW**	A	*CH*	AL	58806		58839
165 007	**NW**	A	*CH*	AL	58807		58840
165 008	**NW**	A	*CH*	AL	58808		58841
165 009	**NW**	A	*CH*	AL	58809		58842
165 010	**NW**	A	*CH*	AL	58810		58843
165 011	**NW**	A	*CH*	AL	58811		58844
165 012	**NW**	A	*CH*	AL	58812		58845
165 013	**NW**	A	*CH*	AL	58813		58846
165 014	**NW**	A	*CH*	AL	58814		58847
165 015	**NW**	A	*CH*	AL	58815		58848
165 016	**NW**	A	*CH*	AL	58816		58849
165 017	**NW**	A	*CH*	AL	58817		58850
165 018	**NW**	A	*CH*	AL	58818		58851
165 019	**NW**	A	*CH*	AL	58819		58852
165 020	**NW**	A	*CH*	AL	58820		58853
165 021	**NW**	A	*CH*	AL	58821		58854
165 022	**NW**	A	*CH*	AL	58822		58855
165 023	**NW**	A	*CH*	AL	58873		58867
165 024	**NW**	A	*CH*	AL	58874		58868
165 025	**NW**	A	*CH*	AL	58875		58869
165 026	**NW**	A	*CH*	AL	58876		58870
165 027	**NW**	A	*CH*	AL	58877		58871
165 028	**NW**	A	*CH*	AL	58878		58872
165 029	**NW**	A	*CH*	AL	58823	55404	58856
165 030	**NW**	A	*CH*	AL	58824	55405	58857
165 031	**NW**	A	*CH*	AL	58825	55406	58858
165 032	**NW**	A	*CH*	AL	58826	55407	58859
165 033	**NW**	A	*CH*	AL	58827	55408	58860
165 034	**NW**	A	*CH*	AL	58828	55409	58861
165 035	**NW**	A	*CH*	AL	58829	55410	58862
165 036	**NW**	A	*CH*	AL	58830	55411	58863
165 037	**NW**	A	*CH*	AL	58831	55412	58864
165 038	**NW**	A	*CH*	AL	58832	55413	58865
165 039	**NW**	A	*CH*	AL	58833	55414	58866

CLASS 165/1   BREL NETWORK TURBO

DMCL–DMS or DMCL–MS–DMS. Built for Thames Trains services.

Engines: One Perkins 2006-TWH of 260 kW (350 h.p.) per car.
Bogies: One BREL P3 and one BREL T3 per car.
Transmission: Hydraulic. Voith T211r with Gmeinder final drive.
Gangways: Within unit only.
Doors: Sliding plug.
Accommodation: 2+3 facing/unidirectional (standard class), 2+2 facing (first class).
Dimensions: 23.50 x 2.85 m.
Maximum Speed: 90 m.p.h.

58953–58969. DMCL. Dia. DP320. Lot No. 31098 York 1992. 24/60 1T. 37.0 t.
58879–58898. DMCL. Dia. DP319. Lot No. 31096 York 1992. 16/72 1T. 37.0 t.
MS. Dia. DR208. Lot No. 31099 York 1992. –/106. 37.0 t.
DMS. Dia. DP253. Lot No. 31097 York 1992. –/98. 37.0 t.

165 101	**NW**	A	*TT*	RG	58916	55415	58953
165 102	**NW**	A	*TT*	RG	58917	55416	58954
165 103	**NW**	A	*TT*	RG	58918	55417	58955
165 104	**NW**	A	*TT*	RG	58919	55418	58956
165 105	**NW**	A	*TT*	RG	58920	55419	58957
165 106	**NW**	A	*TT*	RG	58921	55420	58958
165 107	**NW**	A	*TT*	RG	58922	55421	58959
165 108	**NW**	A	*TT*	RG	58923	55422	58960
165 109	**NW**	A	*TT*	RG	58924	55423	58961
165 110	**NW**	A	*TT*	RG	58925	55424	58962
165 111	**NW**	A	*TT*	RG	58926	55425	58963
165 112	**NW**	A	*TT*	RG	58927	55426	58964
165 113	**NW**	A	*TT*	RG	58928	55427	58965
165 114	**NW**	A	*TT*	RG	58929	55428	58966
165 115	**NW**	A	*TT*	RG	58930	55429	58967
165 116	**NW**	A	*TT*	RG	58931	55430	58968
165 117	**NW**	A	*TT*	RG	58932	55431	58969
165 118	**NW**	A	*TT*	RG	58879		58933
165 119	**NW**	A	*TT*	RG	58880		58934
165 120	**NW**	A	*TT*	RG	58881		58935
165 121	**NW**	A	*TT*	RG	58882		58936
165 122	**NW**	A	*TT*	RG	58883		58937
165 123	**NW**	A	*TT*	RG	58884		58938
165 124	**NW**	A	*TT*	RG	58885		58939
165 125	**NW**	A	*TT*	RG	58886		58940
165 126	**NW**	A	*TT*	RG	58887		58941
165 127	**NW**	A	*TT*	RG	58888		58942
165 128	**NW**	A	*TT*	RG	58889		58943
165 129	**NW**	A	*TT*	RG	58890		58944
165 130	**NW**	A	*TT*	RG	58891		58945
165 131	**NW**	A	*TT*	RG	58892		58946
165 132	**NW**	A	*TT*	RG	58893		58947
165 133	**NW**	A	*TT*	RG	58894		58948
165 134	**NW**	A	*TT*	RG	58895		58949
165 135	**NW**	A	*TT*	RG	58896		58950
165 136	**NW**	A	*TT*	RG	58897		58951
165 137	**NW**	A	*TT*	RG	58898		58952

CLASS 166   ABB NETWORK EXPRESS TURBO

DMCL (A)–MS–DMCL (B). Built for Paddington–Oxford/Newbury services. Air conditioned.

Engines: One Perkins 2006-TWH of 260 kW (350 h.p.) per car.
Bogies: One BREL P3 and one BREL T3 per car.
Transmission: Hydraulic. Voith T211r with Gmeinder final drive.
Gangways: Within unit only.
Doors: Sliding plug.
Accommodation: 2+3 facing/unidirectional (standard class) with 20 standard class seats in 2+2 format in DMCL(B), 2+2 facing (first class).
Dimensions: 22.91 x 2.81 m (DMCL), 22.72 x 2.81 m (MS).
Maximum Speed: 90 m.p.h.

DMCL (A). Dia. DP321. Lot No. 31116 York 1992–3. 16/75 1T. 40.62 t.
MS. Dia. DR209. Lot No. 31117 York 1992–3. –/96. 38.04 t.
DMCL (B). Dia. DP321. Lot No. 31116 York 1992–3. 16/72 1T. 40.64 t.

166 201	**NW**	A	*TT*	RG	58101	58601	58122
166 202	**NW**	A	*TT*	RG	58102	58602	58123
166 203	**NW**	A	*TT*	RG	58103	58603	58124
166 204	**NW**	A	*TT*	RG	58104	58604	58125
166 205	**NW**	A	*TT*	RG	58105	58605	58126
166 206	**NW**	A	*TT*	RG	58106	58606	58127
166 207	**NW**	A	*TT*	RG	58107	58607	58128
166 208	**NW**	A	*TT*	RG	58108	58608	58129
166 209	**NW**	A	*TT*	RG	58109	58609	58130
166 210	**NW**	A	*TT*	RG	58110	58610	58131
166 211	**NW**	A	*TT*	RG	58111	58611	58132
166 212	**NW**	A	*TT*	RG	58112	58612	58133
166 213	**NW**	A	*TT*	RG	58113	58613	58134
166 214	**NW**	A	*TT*	RG	58114	58614	58135
166 215	**NW**	A	*TT*	RG	58115	58615	58136
166 216	**NW**	A	*TT*	RG	58116	58616	58137
166 217	**NW**	A	*TT*	RG	58117	58617	58138
166 218	**NW**	A	*TT*	RG	58118	58618	58139
166 219	**NW**	A	*TT*	RG	58119	58619	58140
166 220	**NW**	A	*TT*	RG	58120	58620	58141
166 221	**NW**	A	*TT*	RG	58121	58621	58142

CLASS 168   ADTRANZ TURBOSTAR

DMSL (A)–MSL–MS–DMSL (B). New units under construction for Chiltern Railways. Aluminium bodies. Air conditioned.

Engines: One MTU 6R183TD13H of 315 kW (422 h.p.) at 1900 r.p.m. per car.
Bogies: One ADtranz P3–23 and one BREL T3–23 per car.
Transmission: Hydraulic. Voith T211rzze to ZF final drive.
Gangways: Within unit only.
Doors: Swing plug.
Accommodation: 2+2 facing/unidirectional.

Dimensions: 23.00 x 2.7 m (DMSL), 22.8 x 2.7 m (MS).
Maximum Speed: 100 m.p.h.

DMSL (A). Dia. DP2 . ADtranz Derby 1998. –/56 + 4 tip-up 1TD.
MSL. Dia. DR2 . ADtranz Derby 1998. –/72 1T.
MS. Dia. DR2 . ADtranz Derby 1998. –/76 1T.
DMSL (B). Dia. DP2 . ADtranz Derby 1998. –/72 1T. Catering point.

168 001	**CI**	P	*CH*	58151	58651	58451	58251
168 002	**CI**	P	*CH*	58152	58652	58452	58252
168 003	**CI**	P	*CH*	58153	58653	58453	58253
168 004	**CI**	P	*CH*	58154	58654	58454	58254
168 005	**CI**	P	*CH*	58155	58655	58455	58255

CLASS 170   ADTRANZ TURBOSTAR

Various formations. New units under construction. Aluminium bodies. Air
conditioned.

Engines: One MTU 6R183TD13H of 315 kW (422 h.p.) at 1900 r.p.m. per car.
Bogies: One ADtranz P3–23 and one BREL T3–23 per car.
Transmission: Hydraulic. Voith T211rzze to ZF final drive.
Gangways: Within unit only.
Doors: Swing plug.
Accommodation: 2+2 facing/unidirectional.
Dimensions: 23.00 x 2.7 m (DMSL), 22.8 x 2.7 m (MS).
Maximum Speed: 100 m.p.h.

Note: Individual car numbers not yet allocated due to review of numbering
systems.

Class 170/1. DMCL(A)–DMCL(B). Vehicles for Midland Main Line.

DMCL (A). Dia. DP3 . ADtranz Derby 1998. 12/45 1TD.
DMCL (B). Dia. DP3 . ADtranz Derby 1998. 12/52 1T. Catering point.

170 101	**MM**	P	*ML*
170 102	**MM**	P	*ML*
170 103	**MM**	P	*ML*
170 104	**MM**	P	*ML*
170 105	**MM**	P	*ML*
170 106	**MM**	P	*ML*
170 107	**MM**	P	*ML*
170 108	**MM**	P	*ML*
170 109	**MM**	P	*ML*
170 110	**MM**	P	*ML*
170 111	**MM**	P	*ML*
170 112	**MM**	P	*ML*
170 113	**MM**	P	*ML*
170 114	**MM**	P	*ML*
170 115	**MM**	P	*ML*
170 116	**MM**	P	*ML*
170 117	**MM**	P	*ML*

Class 170/2. DMCL–MS–DMSL.

DMCL. Dia. DP3   . ADtranz Derby 1998. 30/5 1TD.
MSL. Dia. DR2   . ADtranz Derby 1998. –/65 1T. Catering point.
DMSL. Dia. DP2   . ADtranz Derby 1998. –/70 1T.

170 201	P
170 202	P
170 203	P
170 204	P
170 205	P
170 206	P
170 207	P
170 208	P

Class 170/3. Vehicles for Porterbrook. Details to be specified by customers.

170 301	P
170 302	P
170 303	P
170 304	P
170 305	P
170 306	P
170 307	P
170 308	P
170 309	P
170 310	P
170 311	P
170 312	P
170 313	P
170 314	P
170 315	P
170 316	P
170 317	P
170 318	P
170 319	P
170 320	P
170 321	P
170 322	P

Class 170/4. DMCL–MS–DMSL. Vehicles for a TOC (not yet confirmed).

170 401	P
170 402	P
170 403	P
170 404	P
170 405	P
170 406	P
170 407	P
170 408	P
170 409	P

3. DIESEL ELECTRIC MULTIPLE UNITS

All ex BR Southern Region diesel-electric multiple unit power cars have above-floor-mounted engines and all vehicles are equipped with buckeye couplings and were built at Eastleigh with frames laid at Ashford.

CLASS 201/202 PRESERVED 'HASTINGS' UNIT

DMBSO–3TSOL–DMBSO.
Preserved unit made up from 2 Class 201 short-frame cars and 2 Class 202 long-frame cars. The 'Hastings' units were made with narrow body-profiles for use on the section between Tonbridge and Battle which had tunnels of restricted loading gauge. These tunnels were converted to single track operation in the 1980s thus allowing standard loading gauge stock to be used. The set also contains a Class 411 EMU trailer (not Hastings line gauge).

Engine: English Electric 4SRKT engines of 370 kW (500 h.p.).
Transmission: Two EE 507 traction motors on the inner bogie.
Gangways: Within unit only.
Dimensions: 17.68 x 2.50 m (60000/60501), 19.66 x 2.50 m. (60118/60529).
19.75 x 2.82 m (70262).
Maximum Speed: 75 m.p.h.

60000. DMBSO. Dia DB203. Lot No. 30329 1957. –/22. 54 t.
60118. DMBSO. Dia DB203. Lot No. 30395 1957. –/30. 55 t. Renumbered from 60018.
60501. TSOL. Dia DB204. Lot No. 30331 1957. –/52 2T. 29 t.
60529. TSOL. Dia DH203. Lot No. 30397 1957. –/60 2T. 30 t.
70262. TSOL (ex Class 411/5 EMU). Dia. EH282. Lot No. 30455 1958–9. –/64 2T. 33.78 t.

201 001   **SG**   HD   *SS*   SE   60000 60501 70262 60529 60118

60000 is named 'HASTINGS'.

CLASS 205/0   3H

DMBSO–TSO–DTCsoL or DMBSO–DTCsoL.

Engine: English Electric 4SRKT engines of 450 kW (600 h.p.).
Transmission: Two EE 507 traction motors on the inner bogie.
Gangways: Non-gangwayed.
Dimensions: 20.28 x 2.82 m.
Maximum Speed: 75 m.p.h.

60108–117/154. DMBSO. Dia DB203. Lot No. 30332 1957. –/52. 56 t.
60122–124. DMBSO. Dia DB203. Lot No. 30540 1958–59. –/52. 56 t.
60146–151. DMBSO. Dia DB204. Lot No. 30671 1960–62. –/42. 56 t.
60650–670. TSO. Dia DH203. Lot No. 30542 1958–59. –/104. 30 t.
60673–678. TSO. Dia DH203. Lot No. 30672 1960–62. –/104. 30 t.
60800–811. DTCsoL. Dia DE302. Lot No. 30333 1956–57. 19/50 2T. 32 t.

60822–824. DTCsoL. Dia DE302. Lot No. 30541 1958–59. 19/50 2T. 32 t.
60827–832. DTCsoL. Dia DE303. Lot No. 30673 1960–62. 13/62 2T. 32 t.

§ One compartment of DTCsoL converted to luggage compartment. 13/50 2T. Dia. DE301.

Notes: 60154 was renumbered from 60100.

205 001	§	**N**	P	*SC*	SU	60154	60650	60800
205 009		**N**	P	*SC*	SU	60108	60658	60808
205 012		**N**	P	*SC*	SU	60111	60661	60811
205 018		**N**	P	*SC*	SU	60117	60674	60828
205 023		**N**	P		ZG	60122		60822
205 024	§	**N**	P	*SC*	SU	60123	60669	60823
205 025	§	**N**	P	*SC*	SU	60124	60670	60824
205 028		**CX**	P	*SC*	SU	60146	60673	60827
205 032		**N**	P	*SC*	SU	60150	60677	60831
205 033		**CX**	P	*SC*	SU	60151	60678	60832
Spare		**N**	P		SE		60664	
Spare		**N**	P		SE		60665	
Spare		**N**	P		SE		60668	

CLASS 205/1   3H

DMBSO–TSOL–DTSOL. Refurbished 1980. Fluorescent lighting. PA.

Engine: English Electric 4SRKT engines of 450 kW (600 h.p.).
Transmission: Two EE 507 traction motors on the inner bogie.
Gangways: Within unit only.
Dimensions: 20.28 x 2.82 m.
Maximum Speed: 75 m.p.h.

DMBSO. Dia DB203. Lot No. 30332 1957. –/39. 57 t.
TSOL (ex Class 411/5 EMU). Dia. EH282. Converted from loco-hauled TSO 4059 Lot No. 30149 Ashford/Swindon 1955–7. –/64 2T. 33.78 t.
DTSOL. Dia DE204. Lot No. 30333 1957. –/76 2T. 32 t.

205 205		**N**	P	*SC*	SU	60110	71634	60810

CLASS 207/0   2D

DMBSO–DTSO (formerly DMBSO–TCsoL–DTSO).
These units were built for the Oxted line and therefore referred to as 'Oxted' units. They were made with a narrower body-profile which also allowed them to be used through the restricted loading-gauge Somerhill Tunnel between Tonbridge and Grove Junction (Tunbridge Wells). This tunnel was converted to single track operation in the 1980s thus allowing standard loading gauge stock to be used.

Engine: English Electric 4SRKT engines of 450 kW (600 h.p.).
Transmission: Two EE 507 traction motors on the inner bogie.
Gangways: Non-gangwayed.

Dimensions: 20.34 x 2.74 m. (DMBSO), 20.32 x 2.74 m. (DTSO), 20.34 x 2.74 m. (TCsoL).
Maximum Speed: 75 m.p.h.

DMBSO. Dia DB205. Lot No. 30625 1962. –/42. 56 t.
TCsoL. Dia DH301. Lot No. 30626 1962. 24/42 1T. 31 t.
DTSO. Dia DE201. Lot No. 30627 1962. –/76. 32 t.

207 017	**N**	P	*SC*	SU	60142	60916
Spare	**N**	P		SE	60135 60616	
Spare	**N**	P		ZG	60138	

CLASS 207/1   3D

DMBSO–TSOL–DTSO.
Gangwayed sets with a Class 411 EMU trailer in the centre.

Engine: English Electric 4SRKT engines of 450 kW (600 h.p.).
Transmission: Two EE 507 traction motors on the inner bogie.
Gangways: Fitted with gangways within unit.
Dimensions: 20.34 x 2.74 m. (DMBSO), 20.32 x 2.74 m. (DTSO).
Maximum Speed: 75 m.p.h.

DMBSO. Dia DB205. Lot No. 30625 1962. –/40. 56 t.
70286. TSOL (ex Class 411/5 EMU). Dia. EH282. Lot No. 30455 1958–9. –/64 2T. 33.78 t.
70547/9. TSOL (ex Class 411/5 EMU). Dia. EH282. Lot No. 30620 1960–61 – /64 2T. 33.78 t.
DTSO. Dia DE201. Lot No. 30627 1962. –/75. 32 t.

207 201	**N**	P	*SC*	SU	60129	70286	60903	Ashford Fayre
207 202	**N**	P	*SC*	SU	60130	70549	60904	Brighton Royal Pavilion
207 203	**N**	P	*SC*	SU	60127	70547	60901	

4. SERVICE DMUs

This section contains service vehicles, i.e. vehicles not used for the carrying of passengers which are numbered or renumbered in the special service stock number series or in the internal user series. The last capital stock number carried is shown in parentheses.

LABORATORY COACH DERBY LIGHTWEIGHT

Laboratory Coach 19. Converted from a Derby Lightweight single unit.

Engines: Two BUT of 112 kW (150 hp).
Transmission: Mechanical. Cardan shaft and freewheel to a four-speed epicyclic gearbox with a further cardan shaft to the final drive, each engine driving the inner axle of one bogie.
Gangways: Non gangwayed single car with cab at each end.
Dimensions: 17.53 x 2.82 m.

DMBS. Lot No. 30380 Derby 1957.

G	SO	*RS*	ZA	975010	(79900)	Iris

CLASS 101   METRO-CAMMELL

DMBS–DMCL or DMBS–DMBS or DTCL. For details see page 10. Converted 1990 (i,l), 1993 (s).

51427. DMBS. Dia. DQ202. Lot No. 30500 1959. 32.5 t.
53193. DMCL. Dia. DP317. Lot No. 30256 1957. 32.5 t.
53200–53208. DMBS. Dia. DQ202. Lot No. 30259 1957. 32.5 t.
53222–53231. DMBS. Dia. DQ202. Lot No. 30261 1957. 32.5 t.
53291. DMBS. Dia. DQ202. Lot No. 30270 1957. 32.5 t.
53308. DMBS. Dia. DQ202. Lot No. 30275 1957. 32.5 t.
53321–53338. DMCL. Dia. DP317. Lot No. 30276 1958. 32.5 t.
54342. DTCL. Dia. DS302. Lot No. 30468 1959. 25.5 t.

Non-standard livery: Grey.

i – Internal user (stores van).
† – Lab Coach 19. Iris 2.

–	† **0**	SO	*TE*	ZA	977693	(53222)	977694	(53338)
960 991	**N**	RT	*SA*	LO	977895	(53308)	977896	(53331)
960 992		RT	*SA*	LO	977897	(53203)	977898	(53193)
960 993		RT	*SA*	LO	977899	(51427)	977900	(53321)
960 994		RT	*SA*	LO	977901	(53200)	977902	(53231)
960 995		RT	*SA*	LO	977903	(53208)	977904	(53291)
–	i	ML	*ST*	NL	042222	(54342)		

CLASS 101 TRACTOR UNIT

Converted 1986 from Class 101 DMBS's. For details of Class 101 see page 10.

51433. DMBS. Lot No. 30500 Metro-Cammell 1959. 32.5 t.
53167. DMBS. Lot No. 30254 Metro-Cammell 1957. 32.5 t.

Non-standard livery: Serco. Grey with broad diagonal red stripe.

–	**0**	SO	*TE*	ZA	977391 (51433) 977392 (53167)

CLASS 114 DERBY HEAVYWEIGHT

Converted 1992. Used as a route learning unit.

Engines: Two Leyland TL11 of 152 kW (205 hp) per power car.
Transmission: Mechanical. Cardan shaft and freewheel to a four-speed epicylic gearbox with a further cardan shaft to the final drive, each engine driving the inner axle of one bogie.
Gangways: Midland scissors type. Within unit only.
Doors: Slam.
Bogies: DD9 (motor) and DT9c (trailer).
Dimensions: 20.45 x 2.82 m.

55929. DMPMV. Lot No. 30209 Derby 1958. 29.0 t.
54904. DTPMV. Lot No. 30210 Derby 1958. 29.2 t.

Non-standard livery: Grey, red and yellow.

–	**0**	E	*EW*	CF	977775 (55929) 977776 (54904)

CLASS 121/122 PRESSED STEEL/GLOUCESTER

Sandite cars. For DMBS details see page 11.

55019. Class 122 DMBS. Dia. DX202. Lot No. 30419 Gloucester 1958. 36.5 t.
55020–55028. Class 121 DMBS. Dia. DX201. Lot No. 30518 Pressed Steel 1960. 38.0 t.

960 002	**N**	RT	*SA*	RG	977722	(55020)
960 010	**N**	RT	*SA*	RG	977858	(55024)
960 011	**N**	RT	*SA*	LO	977859	(55025)
960 012	**N**	RT	*SA*	RG	977873	(55022)
960 013	**N**	RT	*SA*	RG	977866	(55030)
960 014	**N**	RT	*SA*	RG	977860	(55028)
960 015	**RT**	RT	*SA*	BY	975042	(55019)
960 021	**N**	RT	*SA*	BY	977723	(55021)

CLASS 205   TRACTOR UNIT

DMBSO–DMBSO. Converted 1994. For details see pages 48.

930 301   RT   SU   977939 (60145)   977940 (60149)

CLASS 205   SANDITE COACH

TSO. Converted 1994. For details see pages 48. Works with 930 301.

              RT   *SA*   SU   977870 (60660)

TRACK RECORDING UNIT

Built new 1987. Class 150 derivative.

Non-standard livery: Blue and white with a red stripe below windows.

 –   **0**   SO   *TE*   ZA   999600   999601

West Yorkshire PTE liveried Class 158/9 No. 158 902 passes New Barnetby with the 14.28 Cleethorpes–Manchester Airport on 30th May 1997.
Brian Denton

A six-car Class 159 formation led by unit No. 159 017 passes Micheldever with the diverted 08.13 Basingstoke–Paignton service on 14th September 1997. The units are in Network SouthEast livery with South West Trains branding.

Nic Joynson

▲ Class 165/0 No. 165 003 at Reading on 26th August 1997.   **Dave McAlone**

▼ Class 166 No. 166 213 passes Buckland with the 14.03 Gatwick Airport–Reading on 4th October 1997.   **Alex Dasi-Sutton**

▲ The first car to be completed of the new Class 168 'Turbostar' units.
ADtranz

▼ Hastings Diesels owned Class 201 No. 1001 approaches Edenbridge Town with the 15.30 Uckfield–Oxted on the occasion of the Uckfield line gala weekend. The centre car is from a Class 411/5 EMU.
David Brown

▲ Connex liveried Class 205 No. 205 028 works the 17.00 Uckfield–Oxted near Edenbridge Town on 23rd August 1997. **Chris Wilson**

▼ Class 207/2 No. 207 201 'Ashford Fayre' departs from Ashford with the 08.54 service to Eastbourne on 10th April 1997. Note the centre car from a Class 411/5 EMU. **David Brown**

Eurotunnel shuttle loco No. 9002 'STUART BURROWS' arrives at Cheriton terminal with a car carrying train from France. The date is 6th September 1996.

Hugh Ballantyne

▲ Docklands Light Railway Class B92 cars, Nos. 80 and 53 enter Crossharbour station with an Island Gardens service on 24th May 1997. **Peter Fox**

▼ Tyne & Wear Metro car No. 4034 at West Jesmond with an Airport–South Shields service on 25th October 1997. **Peter Fox**

South Yorkshire Supertram car No. 16 at Birley Moor Road on 1st January 1997. Trams and trains were the only public transport operating in Sheffield on New Years Day, as no buses ran.

Peter Fox

5. DMUs AWAITING DISPOSAL

The following withdrawn DMUs are awaiting disposal with the last known storage location shown.

Ex-CAPITAL STOCK

51340	ZH	55303	ZA
51359	BY	55402	ZA
51361	Kineton	55403	ZA
51368	Kineton	55709	NH
54350	Crewe Brook Sidings	59228	Crewe Brook Sidings
55202	ZA	59518	OM
55203	ZA	60200	ZG
55302	ZA	60201	ZG

Ex-SERVICE STOCK

975023	(55001)	LO
975025	(60755)	SL
977191	(56106)	Crewe Brook Sidings
977554	(54182)	Buxton LIP
977696	(60522)	EH
977697	(60523)	ZG
977698	(60152)	ZG
977699	(60153)	ZG

6. UK LIGHT RAIL SYSTEMS & METROS

6.1. BLACKPOOL & FLEETWOOD TRAMWAY

System: 660 V d.c. overhead.
Depot: Rigby Road.
Livery: Cream and green. (many in advertising livery).
Note: Numbers in brackets are pre-1968 numbers.

ONE-MAN CARS

Rebuilt 1972–76 from English Electric railcoaches built 1934–5. Radio fitted.
13 converted (1–13).
Seats: 48.
Traction Motors: Two EE305 of 40 kW.

Note: First numbers in brackets are post 1968 numbers prior to conversion.

5	(609, 221)(U)	11	(615, 268)

OPEN BOAT CARS

Built 1934–5 by English Electric. 12 built (225–236).
Seats: 56.
Traction Motors: Two EE327 of 30 kW.

600*	(225)	604§	(230)	606b	(235)
602†	(227)	605	(233)	607	(236)

† Yellow and black livery.
§ Red and white livery.
b Blue & yellow livery.

REPLICA VANGUARD

Built 1987 on underframe of one man car No. 7.(619–282).
Seats: .
Traction Motors: Two EE327 of 30 kW.

619

BRUSH RAILCOACHES

Built 1937 by Brush. 20 built (284–303).
Seats: 48.
Traction Motors: Two EE305 of 40 kW. (EE327 of 30 kW*).

621	(284)	626	(289)	631	(294)	634	(297)
622*	(285)	627	(290)	632	(295)	636	(299)
623	(286)	630	(293)	633	(296)	637	(300)
625	(288)						

CENTENARY CLASS

Built 1984–7. Body by East Lancs. Coachbuilders, Blackburn. One man operated. Radio fitted.
Seats: 52.
Traction Motors: Two EE305 of 40 kW.

* Rebuilt from GEC car 651.

641	643	645	647
642	644	646	648*

CORONATION CLASS

Built 1953 by Charles Roberts & Co. Resilient wheels. 25 built (304–328).
Seats: 56.
Traction motors: Four Crompton-Parkinson 92 of 34 kW.

660   (324)

PROGRESS TWIN CARS

Motor cars (671–677) rebuilt 1958–60 from English Electric railcoaches.
Seats: 53.
Traction Motors: Two EE305 of 40 kW.
Driving trailers (681–687) built 1960 by Metro-Cammell.
Seats: 53.

671+681	(281+T1)	674+684	(284+T4)	676+686	(286+T6)
672+682	(282+T2)	675+685	(285+T5)	677+687	(287+T7)
673+683	(283+T1)				

SINGLE CARS

Rebuilt 1958–60 from English Electric railcoaches. Originally ran with trailers.
Seats: 48.
Traction Motors: Two EE305 of 40 kW.

678   (278) | 679   (279) | 680   (280)

"BALLOON" DOUBLE DECKERS

Built 1934–5 by English Electric. 700–712 were originally built with open tops, and 706 has now reverted to that condition and is named 'PRINCESS ALICE'.

Seats: 94.
Traction Motors: Two EE305 of 40 kW.

* Converted to ice cream tram seating 64 with an ice cream sales area in one of the lower saloons.
§ Red and white livery.

700	(237)	709	(246)	718	(255)
701§	(238)	710	(247)	719	(256)
702	(239)	711	(248)	720	(257)
703	(240)	712	(249)	721	(258)
704	(241)	713	(250)	722	(259)
706	(243)	715	(252)	723	(260)
707	(244)	716	(253)	724	(261)
708	(245)	717	(254)	726	(263)

ILLUMINATED CARS

732	(168)	Rocket	Seats: 47
733	(209)	Western Train loco. & tender	Seats: 35
734	(174)	Western Train coach	Seats: 60
735	(222)	Hovertram	Seats: 99
736	(170)	HMS Blackpool	Seats: 71

WORKS CARS

259	(748, 624)	PW gang towing car.
260	(751, 628, 291)	Crane car and rail carrier.
749	(S)	Tower wagon trailer.
750		Cable drum trailer.
752	(2, 1)	Rail grinder and snowplough.
754		New works car (unnumbered).

JUBILEE CLASS DOUBLE DECKERS

Rebuilt 1979/82 from Balloon cars. Standard bus ends, thyristor control and stairs at each end. 761 has one door per side whereas 762 has two. Radio fitted.
Seats: 100.
Traction Motors: Two EE305 of 40 kW.

761　(725, 262)　|　762　(714, 251)

PRESERVED CARS

Blackpool & Fleetwood 40	Box car. Bogie single decker built 1914
Bolton 66	Bogie double-decker built 1901

6.2. DOCKLANDS LIGHT RAILWAY

This is a light rail line running in London's East End from Bank, Tower Gateway and Stratford to Island Gardens and Beckton. It is being extended to Lewisham. Originally owned by London Transport, it is now owned by the London Docklands Development Corporation.

System: 750 V d.c. third rail (bottom contact).
Depots: Poplar, Beckton.

CLASS P89   B–2–B

Built 1990 by BREL Ltd. York Works. 28.80 x 2.65 m. Sliding doors. Chopper control. Scharfenberg Couplers.

Weight: 39 t.
Seats: 84.
Traction Motors: Two GEC of 185 kW.
Max. Speed: 80 km/h.
Electric Brake: Rheostatic.

12(S)	15(S)	18(S)	20
13(S)	16(S)	19(S)	21(S)
14(S)	17(S)		

CLASS B90   B–2–B

Built 1991–2 by BN Construction, Bruges, Belgium. (now Bombardier BN). 28.80 x 2.65 m. Sliding doors. End doors for staff use. Chopper control. Scharfenberg Couplers. These units are to be converted for Seltrack signalling.

Weight: 36 t.
Seats: 66 + 4 tip-up.
Traction Motors: Two Brush of 140 kW.
Max. Speed: 80 km/h.
Electric Brake: Rheostatic.

22	28	34	40
23	29	35	41
24	30	36	42
25	31	37	43
26	32	38	44
27	33	39	

CLASS B92 B–2–B

Built 1992–5 by BN Construction, Bruges, Belgium. (now Bombardier BN).
28.80 x 2.65 m. Sliding doors. End doors for staff use. Chopper control.
Scharfenberg Couplers. Fitted with Seltrack signalling.

Weight: 36 t.
Seats: 66 + 4 tip-up.
Traction Motors: Two Brush of 140 kW.
Max. Speed: 80 km/h.
Electric Brake: Rheostatic.

45	57	69	81
46	58	70	82
47	59	71	83
48	60	72	84
49	61	73	85
50	62	74	86
51	63	75	87
52	64	76	88
53	65	77	89
54	66	78	90
55	67	79	91
56	68	80	

6.3. GREATER MANCHESTER METROLINK

This light rail system runs from Bury to Altrincham through the streets of Manchester, with a spur to Piccadilly. An extension is being built to Salford Quay and Eccles.

System: 750 V d.c. overhead.
Depot: Queens Road.

SIX-AXLE ARTICULATED CARS   Bo–2–Bo

Built 1991–2 by Firema, Italy. Power operated sliding doors. Chopper control. Scharfenberg Couplers.

Weight: 45 t.
Seats: 84.
Dimensions: 29.00 x 2.65 m.
Traction Motors: Four GEC of 130 kW.
Braking: Rheostatic, regenerative, disc and emergency track brakes.

1001	CHILDREN'S HOSPITALS APPEAL 1
1002	
1003	
1004	THE ROBERT OWEN
1005	GREATER ALTRINCHAM ENTERPRISE
1006	
1007	
1008	MANCHESTER AIRPORT
1009	
1010	MANCHESTER CHAMPION
1011	
1012	KERRY
1013	THE FUSILIER
1014	THE CITY OF DRAMA
1015	SPARKY
1016	
1017	
1018	
1019	THE ERIC BLACK
1020	THE DAVID GRAHAM CBE
1021	THE GREATER MANCHESTER RADIO
1022	THE GRAHAM ASHWORTH
1023	
1024	THE JOHN GREENWOOD
1025	
1026	THE POWER

SPECIAL PURPOSE VEHICLE

Built 1991 by RFS Industries, Kilnhurst and Brown Root. Used for shunting and track maintenance. Includes a crane.

Unnumbered.

6.4. MIDLAND METRO

This light rail system is under construction and will operate between Birmingham Snow Hill and Central Wolverhampton. It is due to open on 3rd August

System: 750 V d.c. overhead.
Depot: Wednesbury.

SIX-AXLE ARTICULATED CARS　　　Bo–2–Bo

Built 1998 by Ansaldo Trasporti, Italy. Power operated sliding plug doors. IGBT control. Scharfenberg Couplers.

Weight: 35.6 t.
Seats: 58.
Dimensions: 24.00 x 2.65 m.
Traction Motors: Four .
Max. Speed: 75 km/h (47 m.p.h.).
Braking: Rheostatic, regenerative, disc and emergency track brakes.

01	07	13	19
02	08	14	20
03	09	15	21
04	10	16	22
05	11	17	23
06	12	18	24

Note: Actual fleet numbers not definite at this stage.

6.5. SOUTH YORKSHIRE SUPERTRAM

This light rail system has three lines, to Halfway in the south east of Sheffield with a spur from Gleadless Townend to Herdings, to Middlewood in the north west with a spur from Hillsborough to Malin Bridge and to Meadowhall Interchange in the north east adjacent to the large shopping complex. Because of the severe gradients in Sheffield (up to 1 in 10), all axles are powered on these vehicles. The vehicles are owned by South Yorkshire Light Rail Ltd., a subsidiary of South Yorkshire Passenger Transport Executive, but are mortgaged to Lloyd's Bank, whilst the operating company, South Yorkshire Supertram Ltd. has been leased to Stagecoach Holdings Ltd. for 27 years.

System: 750 V d.c. overhead.
Depot: Nunnery.

EIGHT-AXLE ARTICULATED UNITS   B–B–B–B

Built 1993–4 by Duewag, Düsseldorf, Germany.

Weight: 52 t.
Seats: 88.
Dimensions: 34.75 x 2.65 m.
Traction Motors: Four monomotors.
Braking: Rheostatic, regenerative, disc and emergency track brakes.

01	08	14	20
02	09	15	21
03	10	16	22
04	11	17	23
05	12	18	24
06	13	19	25
07			

FOUR WHEELED WORKS CAR   B

Built 1968 by Reichsbahn Ausbesserungswerke Schöneweide, Berlin, East Germany as single-ended passenger car with electrical equipment by LEW Henningsdorf. Converted 1980 to double-ended works car. Delivered to Sheffield on 7th November 1996. To be converted to a rail grinder.

Weight: .
Dimensions: .
Traction Motors: Two.

721 039-4   (5104, 217 303-7)

6.6. STRATHCLYDE PTE UNDERGROUND

This circular 4' gauge underground line in Glasgow is generally referred to as the "Subway".
System: 750 V d.c. third rail.
Depot: Broomloan.

SINGLE CARS   Bo-Bo

Built 1978–9 by Metro-Cammell. Power-operated sliding doors. 12.58 x 2.34 m.

Seats: 36.
Traction Motors: Two GEC G312AZ of 35.6 kW.

101	110	118	126
102	111	119	127
103	112	120	128
104	113	121	129
105	114	122	130
106	115	123	131
107	116	124	132
108	117	125	133
109			

INTERMEDIATE TRAILERS   2–2

Built 1992 by Hunslet TPL. Power-operated sliding doors. 12.58 x 2.34 m.
Seats: 40.

201	203	205	207
202	204	206	208

6.7. TYNE AND WEAR METRO

System: 1500 V d.c. overhead.
Depot: South Gosforth.

BATTERY/OVERHEAD ELECTRIC LOCOS

Built: 1989–80 by Hunslet, Leeds. BSI couplers.
Traction Motors: Hunslet-Greenbat T9-4P.
Weight: 26 t.

BL1 | BL2 | BL3

SIX-AXLE ARTICULATED UNITS   B–2–B

Built 1976, 1978–81 by Metro-Cammell. 27.80 x 2.65m. BSI couplers.
Weight: 39 t.
Seats: 84 (68 r – refurbished units, 70 p – Prototype refurbished unit).
Traction Motors: Two 187 kW monomotor bogies.
Max. Speed: 80 km/h.

4001		4019r	**R**	4037		4055 r	**R**	4073	
4002		4020r	**R**	4038		4056	**A**	4074 r	**R**
4003 r **R**		4021r	**R**	4039	r **A**	4057		4075 r	**B**
4004 r **G**		4022		4040		4058		4076	
4005 r **R**		4023		4041		4059		4077 r	**R**
4006		4024		4042		4060		4078 r	**R**
4007 r **R**		4025		4043	r **R**	4061 r	**G**	4079	
4008 r **B**		4026r	**A**	4044	r **R**	4062		4080 r	**R**
4009		4027r	**R**	4045	r **A**	4063		4081	
4010 r **R**		4028		4046	r **R**	4064 r	**R**	4082 r	**G**
4011		4029		4047		4065 r	**R**	4083 r	**A**
4012 r **A**		4030r	**R**	4048	r **B**	4066 r	**B**	4084	
4013		4031		4049	r **A**	4067		4085 r	**B**
4014		4032		4050		4068 r	**R**	4086 r	**B**
4015		4033r	**B**	4051	r **R**	4069		4087 p	**A**
4016 r **B**		4034r	**R**	4052		4070 r	**R**	4088 r	**R**
4017 r **R**		4035r	**B**	4053		4071		4089 r	**R**
4018		4036r	**G**	4054		4072		4090 r	**R**

Names:

4041   HARRY COWANS | 4065   Catherine Cookson

Liveries:

A Advertising livery.
B Blue.
G Green.
R Red.
Standard livery is yellow and white.

LIVERY CODES

All diesel multiple unit vehicles are in the old blue & grey livery unless otherwise indicated. The colour of the lower half of the bodyside is stated first. Please note that although the former Provincial Services sector is now known as Regional Railways, the former name is used for its original liveries.

CC	BR carmine & cream ("Blood & Custard")
CE	Centro (WMPTE) (grey/light blue/white/green)
CI	Chiltern railways (grey and blue with red stripe)
CX	Connex (yellow & white with blue solebar stripe)
GM	New Greater Manchester PTE (dark grey/red/white/light grey)
G	BR DMU green or Southern Region green
LH	Loadhaul (orange and black)
MM	Midland Mainline (grey and green with three orange bodyside stripes)
MT	Merseytravel (yellow/blue/white/yellow)
N	Network SouthEast (grey/white/red/white/blue/white)
NW	Network SouthEast (white/red/white/blue/white)
O	Other livery (non-standard - refer to text)
P	Provincial Services (grey/light blue/white/dark blue)
PR	Provincial Services railbus variant (dark blue/white/light blue)
RE	Regional Railways Express (buff/light grey/dark grey/light grey/buff with dark blue, white and light blue stripes)
RN	As RR but with green stripe under windows
RR	Regional Railways (grey/light blue/white/dark blue with three black and white stripes at end of each light blue band under cabs)
S	Strathclyde PTE (orange and black)
T	Tyne & Wear PTE (yellow blue & white)
Y	West Yorkshire PTE (red and cream)

OWNER AND OPERATION CODES

This book now uses a (generally) logical system of codes instead of the gobbledygook codes of the BR Rolling Stock Library (RSL). We have decided to do this since RSL information is not officially available to the general public these days and a system of coding which is fairly obvious to the reader is preferred. For passenger train operating companies these are generally based on those used by Railtrack in the Great Britain passenger timetable, but there are a few changes for clarity or to reflect changes since the timetable was printed

OWNER CODES

A	Angel Trains Contracts
E	English Welsh & Scottish Railway
O	Other private owner (refer to text)
P	Porterbrook Leasing Company
ML	Midland Mainline
RT	Railtrack
SO	Serco Railtest

OPERATION CODES

The two letter operation codes give the use to which the vehicle is at present put. For vehicles in regular use, this is the code for the train operating company For other vehicles the actual type of use is shown. If no operation code is shown then the vehicle is not at present in use.

AR	Anglia Railways
CA	Cardiff Railway Company
CH	Chiltern Trains
CR	Crew training
CT	Central Trains
NE	Regional Railways North East
NL	North London Railways
NW	North West Regional Railways
RS	Research
SA	Sandite spraying
SC	Connex South Central
SR	ScotRail
SS	Special services
ST	Stores vehicle
SW	South West Trains
TE	Test trains
TT	Thames Trains
VX	Virgin Cross Country
WW	Wales and West Passenger Trains

DEPOT CODES

AL	Aylesbury TMD
BY	Bletchley TMD
CF	Cardiff Canton TMD
CK	Corkerhill SD (Glasgow)
HA	Haymarket TMD (Edinburgh)
HT	Heaton T&RSMD (Newcastle)
IL	Laira T&RSMD (Plymouth)
IS	Inverness T&RSMD
KI	MoD Kineton (unofficial code)
LO	Longsight TMD (D) (Manchester)
NC	Norwich Crown Point T&RSMD
NH	Newton Heath T&RSMD (Manchester)
NL	Neville Hill T&RSMD (Leeds)
RG	Reading TMD
SA	Salisbury TMD
SE	St. Leonards Railway Engineering Co.
SU	Selhurst TMD (London)
TE	Thornaby TMD
TS	Tyseley TMD (Birmingham)

DEPOT TYPE CODES

T&RSMD	Traction and Rolling Stock Maintenance Depot.
TMD	Traction Maintenance Depot.
TMD (D)	Traction Maintenance Depot (Diesel).
SD	Servicing Depot.

WORKS CODES

ZA	Railway Technical Centre, Derby
ZB	RFS (E) Ltd., Doncaster
ZC	ADtranz Crewe Works
ZD	ADtranz Derby Carriage Works
ZF	ADtranz Doncaster Works
ZG	Wessex Traincare Ltd., Eastleigh Works
ZH	Railcare Ltd., Springburn Works, Glasgow
ZI	ADtranz Ilford Works
ZK	Hunslet-Barclay Ltd., Kilmarnock Works
ZN	Railcare Ltd., Wolverton Works
ZT	ADtranz Trafford Park Wheel Works, Manchester (unofficial code)